W9-AHP-574

Springer Series in
OPTICAL SCIENCES 139

founded by H.K.V. Lotsch

Springer Series in
OPTICAL SCIENCES

The Springer Series in Optical Sciences, under the leadership of Editor-in-Chief *William T. Rhodes*, Georgia Institute of Technology, USA, provides an expanding selection of research monographs in all major areas of optics: lasers and quantum optics, ultrafast phenomena, optical spectroscopy techniques, optoelectronics, quantum information, information optics, applied laser technology, industrial applications, and other topics of contemporary interest.
With this broad coverage of topics, the series is of use to all research scientists and engineers who need up-to-date reference books.
The editors encourage prospective authors to correspond with them in advance of submitting a manuscript. Submission of manuscripts should be made to the Editor-in-Chief or one of the Editors. See also www.springer.com/series/624

Editor-in-Chief
William T. Rhodes
Georgia Institute of Technology
School of Electrical and Computer Engineering
Atlanta, GA 30332-0250, USA
E-mail: bill.rhodes@ece.gatech.edu

Editorial Board
Ali Adibi
Georgia Institute of Technology
School of Electrical and Computer Engineering
Atlanta, GA 30332-0250, USA
E-mail: adibi@ee.gatech.edu

Toshimitsu Asakura
Hokkai-Gakuen University
Faculty of Engineering
1-I, Minami-26, Nishi 11, Chuo-ku
Sapporo, Hokkaido 064-0926, Japan
E-mail: asakura@eli.hokkai-s-u.ac.jp

Theodor W. Hansch
Max-Planck-Institut für Quantenoptik
Hans-Kopfermann-Straße I
85748 Garching, Germany
E-mail: t.w.haensch@physik.uni-muenchen.de

Takeshi Kamiya
Ministry of Education, Culture, Sports
Science and Technology
National Institution for Academic Degrees
3-29-1 Otsuka, Bunkyo-ku
Tokyo 112-0012, Japan
E-mail: kamiyatk@niad.ac.jp

Ferenc Krausz
Ludwig-Maximilians-Universität München
Lehrstuhl für Experimentelle Physik
Am Coulombwall 1
85748 Garching, Germany
and
Max-Planck-Institut für Quantenoptik
Hans-Kopfermann-Straße 1
85748 Garching, Germany
E-mail: ferenc.krausz@mpq.mpg.de

Bo Monemar
Department of Physics
and Measurement Technology
Materials Science Division
Linkoping University
58183 Linköping, Sweden
E-mail: bom@ifm.liu.se

Herbert Venghaus
Fraunhofer Institut für Nachrichtentechnik
Heinrich-Hertz-Institut
Einsteinufer 37
10587 Berlin, Germany
E-mail: venghaus@hhi.de

Horst Weber
Technische Universität Berlin
Optisches Institut
Straße des 17. Juni 135
10623 Berlin, Germany
E-mail: weber@physik.tu-berlin.de

Harald Weinfurter
Ludwig-Maximilians-Universität München
Sektion Physik
Schellingstraße 4/III
80799 München, Germany
E-mail: harald.weinfurter@physik.uni-muenchen.de

Motoichi Ohtsu
(Ed.)

Progress in Nano-Electro-Optics VI

Nano-Optical Probing,
Manipulation, Analysis,
and Their Theoretical Bases

With 107 Figures

 Springer

Professor Dr. Motoichi Ohtsu
Department of Electronics Engineering
School of Engineering
The University of Tokyo
7-3-1 Hongo, Bunkyo-ku, Tokyo 113-8656, Japan
E-mail: ohtsu@ee.t.u-tokyo.ac.jp

Springer Series in Optical Sciences   ISSN 0342-4111   e-ISSN 1556-1534

ISBN 978-3-540-77894-3   e-ISBN 978-3-540-77895-0

Library of Congress Cataloging-in-Publication Data

Progress in nano-electro-optics VI: nano-optical probing, manipulation, analysis, and their theoretical bases/
Motoichi Ohtsu (ed.). p.cm. – (Springer series in optical sciences; v. 139)
Includes bibliographical references and index.
ISBN 978-3-540-77894-9 (alk. paper)
1. Electrooptics. 2. Nanotechnology. 3. Near-field microscopy. I. Ohtsu, Motoichi. II. Series.
TA1750 .P75 2002   621.381'045-dc21   2002030321

© Springer-Verlag Berlin Heidelberg 2008

Typesetting by the authors and VTEX, using a Springer LATEX macro
Cover concept: eStudio Calamar Steinen
Cover production: WMX Design GmbH, Heidelberg

SPIN: 12216208   57/3180/vtex
Printed on acid-free paper

9 8 7 6 5 4 3 2 1

springer.com

Preface to *Progress in Nano Electro-Optics*

Recent advances in electro-optical systems require dramatic increases in the degree of integration between photonic and electronic devices for large-capacity, ultrahigh-speed signal transmission and information processing. To meet this demand—which will become increasingly strict in the future—device size has to be scaled down to nanometric dimensions. In the case of photonic devices, this requirement cannot be met only by decreasing the material sizes. It is necessary to decrease the size of the electromagnetic field used as a carrier for signal transmission. Such a decrease in the electromagnetic field's size, beyond the diffraction limit of the propagating field, can be realized in optical near fields.

Near-field optics has progressed rapidly in elucidating the science and technology of such fields. Exploiting an essential feature of optical near fields, i.e., the resonant interaction between electromagnetic fields and matter in nanometric regions, important applications and new directions have been realized and significant progress has been reported. These advances have come from studies of spatially resolved spectroscopy, nanofabrication, nanophotonic devices, ultrahigh-density optical memory and atom manipulation. Since nanotechnology for fabricating nanometric materials has progressed simultaneously, combining the products of these studies can open new fields to meet the requirements of future technologies.

This unique monograph series, entitled *Progress in Nano Electro-Optics*, is being introduced to review the results of advanced studies in the field of electro-optics at nanometric scales. The series covers the most recent topics of theoretical and experimental interest on relevant fields of study (e.g., classical and quantum optics, organic and inorganic material science and technology, surface science, spectroscopy, atom manipulation, photonics and electronics). Each chapter is written by leading scientists in the relevant field. Thus, high-quality scientific and technical information is provided to scientists, engineers and students who are and who will be engaged in nano electro-optics and nanophotonics research.

I gratefully thank the members of the editorial advisory board for valuable suggestions and comments on organizing this monograph series. I express my special thanks to Dr. T. Asakura, Editor of the Springer Series in Optical Sciences, Professor Emeritus, Hokkaido University for recommending me to publish this monograph series. Finally, I extend an acknowledgement to Dr. Claus Ascheron of Springer-Verlag, for his guidance and suggestions, and to Dr. H. Ito, an associate editor, for his assistance throughout the preparation of this monograph series.

Yokohama   *Motoichi Ohtsu*
October 2002

Preface to Volume VI

This volume contains five review articles focusing on various, but mutually related topics in nano electro-optics. The first article describes recent developments in near-field optical microscopy and spectroscopy. Owing to a spatial resolution as high as 1–30 nm, spatial profiles of local density of states have been mapped into a real space. This clarifies the fundamental aspects of both localized and delocalized electrons in interface and alloy disorder systems. This kind of study for optical probing and manipulation of electron quantum states in semiconductors at the nanoscale is vital to the development of future nanophotonic devices.

The second article is devoted to describing a quantum theoretical approach to an interacting system of photon, electronic excitation and phonon fields on a nanometer scale—a theoretical basis for nanophotonics. It discusses the phonon's role and localization mechanism of photons in such a system. It allows us not only to understand an elementary process of photochemical reactions with optical near fields, but also to generally explore phonons' roles in nanostructures interacting with localized photon fields.

The third article concerns the visible laser desorption/ionization of bio-molecules from the gold-coated porous silicon, gold nanorod arrays and nanoparticles. Interesting phenomena have been observed to clearly suggest near-field effects on the desorption/ionization mechanism. The techniques presented offer a potential analytical method for the low-molecular weight analytes that are rather difficult to handle in the conventional matrix-assisted laser desorption/ionization (MALDI) mass spectrometry.

The fourth article deals with a near-field optical lithography (NFOL) as an instance of nanofabrication using optical near fields, a method which is not affected by the diffraction limit of light. A bilayer resist process has been developed that enables one to form fine patterns on a structure with a practical aspect ratio. This process was successfully applied to an ultraviolet second harmonic generation (SHG) wavelength conversion device. These technologies are expected to provide a practical fabrication method for optical devices.

The last article reviews recent advances in optical manipulation of nanometric objects using resonant radiation force. Theoretical bases and unified expressions applicable to the different-size regimes—i.e., from the atomic to macroscopic regimes—are presented. According to the theoretical predictions obtained, experimental achievements are described on optical transport of nanoparticles in superfluid ^{4}He, selectively manipulated by the resonant radiation force.

As was the case of volumes I–V, this volume is published with the support of an associate editor and members of editorial advisory board. They are:

Associate editor: Kobayashi, K. (Tokyo Inst. Tech., Japan)

Editorial advisory board: Barbara, P.F. (Univ. of Texas, USA)
                           Bernt, R. (Univ. of Kiel, Germany)
                           Courjon, D. (Univ. de Franche-Comté, France)
                           Hori, H. (Univ. of Yamanashi, Japan)
                           Kawata, S. (Osaka Univ., Japan)
                           Pohl, D. (Univ. of Basel, Switzerland)
                           Tsukada, M. (Waseda Univ., Japan)
                           Zhu, X. (Peking Univ., China)

I hope that this volume will be a valuable resource for readers and for future specialists.

Tokyo   *Motoichi Ohtsu*
April 2008

Contents

List of Contributors

Lee Chuin Chen
Clean Energy Research Center
University of Yamanashi
4-3-11 Takeda, Kofu
Yamanashi 400-8511, Japan
chenleechuin@yahoo.com

Kenzo Hiraoka
Clean Energy Research Center
University of Yamanashi
4-3-11 Takeda, Kofu
Yamanashi 400-8511, Japan
hiraoka@ab11.yamanashi.ac.jp

Hirokazu Hori
Interdisciplinary Graduate School
of Medicine and Engineering
University of Yamanashi
4-3-11 Takeda, Kofu
Yamanashi 400-8551, Japan
hirohori@yamanashi.ac.jp

Takuya Iida
School of Engineering
Osaka Prefecture University
1-1 Gakuen-cho, Naka-ku, Sakai
Osaka 599-8531, Japan
takuya-iida@pe.osakafu-u.ac.jp

Hajime Ishihara
School of Engineering
Osaka Prefecture University
1-1 Gakuen-cho, Naka-ku, Sakai
Osaka 599-8531, Japan
ishi@pe.osakafu-u.ac.jp

Tadashi Kawazoe
School of Engineering
The University of Tokyo
2-11-16 Yayoi, Bunkyo-ku
Tokyo 113-8656, Japan
kawazoe@ee.t.u-tokyo.ac.jp

Kiyoshi Kobayashi
Department of Physics
Tokyo Institute of Technology
2-12-1/H79 O-okayama, Meguro-ku
Tokyo 152-8551, Japan
kkoba@phys.titech.ac.jp

Masayuki Naya
Frontier Core-Technology
Laboratories
Fujifilm Corporation
577 Ushijima, Kaisei-machi,
Ashigarakami-gun
Kanagawa 258-8577, Japan
masayuki_naya@fujifilm.co.jp

xiv   List of Contributors

Motoichi Ohtsu
School of Engineering
The University of Tokyo
2-11-16 Yayoi, Bunkyo-ku
Tokyo 113-8656, Japan
ohtsu@ee.t.u-tokyo.ac.jp

Toshiharu Saiki
Department of Electronics and
Electrical Engineering
Keio University

3-14-1 Hiyoshi, Kohoku-ku,
Yokohama-shi
Kanagawa 223-8522, Japan
saiki@elec.keio.ac.jp

Yuji Tanaka
Department of Physics
Tokyo Institute of Technology
2-12-1 O-okayama, Meguro-ku
Tokyo 152-8551, Japan
tanaka@stat.phys.titech.ac.jp

1

Optical Interaction of Light with Semiconductor Quantum Confined States at the Nanoscale

T. Saiki

1.1 Introduction

Optical probing and manipulation of electron quantum states in semiconductors at the nanoscale are key to developing future nanophotonic devices which are capable of ultrafast and low-power operation [1]. To optimize device performance and to go far beyond conventional devices based on the far-field optics, the degree to which the electron and light are confined must be properly designed and engineered. This is because while stronger confinement of the electron is lets us use its quantum nature, its interaction with light becomes weaker with reduction of the confinement volume. To maximize their interaction, we need the overlap in scale between confinement volume of electron and that of light. More generally, the spatial profile of the light field should be designed to match that of electron wavefunction in terms of phase as well as amplitude.

Semiconductor quantum dots (QDs) provide ideal electron systems because electrons are three-dimensionally confined. This results in a discrete density of states in which the level of energy spacing exceeds the thermal energy. Due to the nature of QDs, they exhibit ultranarrow optical transition spectrum and long duration of coherence [2, 3]. Moreover they can be engineered to have desired properties by controlling the size, shape and strains, as well as by selecting appropriate material. Regarding the size of QDs, with the maturation of crystal growth along with the nanofabrication of semiconductors, we have obtained QDs in a wide rage of sizes from a few nm to larger than 100 nm. For example, interface fluctuation QDs—where excitons by imperfect GaAs quantum are well confined—are extensively studied [4]. By adopting a growth-interruption technique, monolayer-high islands larger than 100 nm develop at the well–barrier interface. Large QDs are advantageous for maximizing the magnitude of the light–electron (exciton) interaction due to the enhancement of oscillator strength, which is proportional to the size of QDs [5].

The progress in light confinement, on the other hand, has also been remarkable [6, 7]. Basically, efforts to focus light more tightly than half the wavelength (diffraction limit) have been motivated by the ultimate spatial resolution of optical microscopy. For example, a near-field scanning optical microscope (NSOM) [6, 7] uses a sharpened optical fiber probe with a small metal hole at its apex to squeeze light in an area determined by the size of the hole. Recent advances in fabrication of NSOM probes enable us to generate a light spot smaller than 10 nm [8]. An optical antenna is also attracting attention due to its higher efficiency in the delivery of energy to a nanofocused spot [9]. Metal nanorods and more sophisticated metal structures provide an opportunity to engineer the light field at the nanoscale with a high degree of freedom.

Broad overlap in the scale between the confinement volume of electrons and light, as described above, leads to changes in their interaction from the far-field counterpart [10]. More specifically, in the case where the spatial resolution of NSOM falls below the size of QD, it becomes possible to directly map out the distribution of the wavefunction [11]. More interestingly, the optical selection rule can be broken; one can excite the dark states whose optical transition is forbidden by the far field and can open new radiative decay channels. The light–matter coupling at the nanoscale offers guiding principles for future nanophotonic devices.

Here, we describe development of a high-resolution NSOM with a carefully designed aperture probe and near-field imaging spectroscopy of quantum confined systems. Thanks to a spatial resolution as high as 1–30 nm, we visualize spatial profiles of local density of states and wavefunctions of electrons confined in QDs and clarify the fundamental aspects of localized and delocalized electrons in interface and alloy disorder systems.

1.2 Near-Field Scanning Optical Microscope

1.2.1 General Description

When a small object is illuminated, its fine structures with high spatial frequency generate a localized field that decays exponentially normal to the object [6, 7]. This evanescent field on the tiny substructure can be used as a local source of light, illuminating and scanning a sample surface so close that the light interacts with the sample without diffraction. A metal opening (aperture) is a popular method for generating a localized optical field suitable for NSOMs. As illustrated in Fig. 1.1, aperture NSOM uses a small opening at the apex of a tapered optical fiber coated with metal. Light sent down the fiber probe and through the aperture illuminates a small area on the sample surface. The fundamental spatial resolution is determined by the diameter of the aperture, which ranges from 10–100 nm.

The simplest setup for imaging spectroscopy based on aperture NSOM is a configuration with local illumination and local collection of light through an aperture, as illustrated in Fig. 1.1. The light emitted by the aperture interacts with the sample locally. Resultant signals from the interaction volume must be collected as efficiently

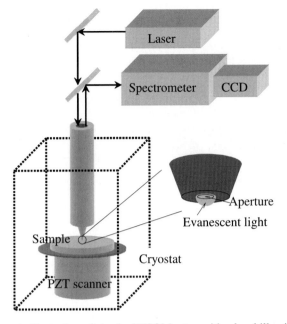

Fig. 1.1. A schematic illustration of standard NSOM setup with a local illumination and local collection configuration

as possible. In photoluminescence (PL) or Raman spectroscopy, the collected signal is dispersed by a spectrometer and is detected by a CCD recording device. The regulation system for tip–sample feedback are essential for NSOM performance, and most NSOMs employ a method similar to that used in an atomic force microscope (AFM), called shear force feedback, the regulation range of which is 0–10 nm [12]. For the measurement at low temperature to reduce phonon-induced broadening, the sample, probe tip, and scanner are placed into a cryostat [13].

1.2.2 Aperture–NSOM Probe

Great effort has been devoted to fabricating the aperture probe, which is the heart of NSOM. Since the quality of the probe determines the spatial resolution and sensitivity of the measurements, tip fabrication remains of major interest in the development of NSOM. To enhance the performance of aperture–NSOM, we focus on two important features of the probe: the light propagation efficiency of the tapered waveguide and the quality of aperture, as illustrated in Fig. 1.2.

Improvement in the optical transmission efficiency (throughput) and collection efficiency of aperture probes is the most important issue to be addressed for the application of NSOM in the spectroscopic studies of nanostructures. The tapered region of the aperture probe operates as a metal-clad optical waveguide. The mode structure in a metallic waveguide is completely different from that in an unperturbed fiber and is characterized by the cutoff diameter and absorption coefficient of the

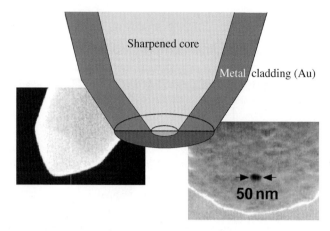

Fig. 1.2. A schematic illustration of aperture–NSOM probe. Scanning electron micrographs of a double-tapered probe (taken prior to metal coating) and a well-defined aperture are also shown

cladding metal. Theoretical and systematic experimental studies have confirmed that the transmission efficiency of the propagating mode decreases in the region where the core diameter is smaller than half the wavelength of the light in the core. The power that is actually delivered to the aperture depends on the distance between the aperture plane and the plane in which the probe diameter is equal to the cutoff diameter; this distance is determined by the taper angle. We therefore proposed a double-tapered structure with a large taper angle [14, 15]. This structure is easily realized using a multistep chemical etching technique, as will be described later. With this technique, the transmission efficiency is much improved by one to two orders of magnitude as compared to the single-tapered probe with a small taper angle.

We used a chemical etching process with buffered HF solution to fabricate the probe. The etching method is easily reproducible and can be used to make many probes at the same time. The details of probe fabrication with selective etching are described in [15]. The taper angle can be adjusted by changing the composition of a buffered HF solution. A two-step etching process is employed to make a double-tapered probe. Another important advantage of the chemical etching method is the excellent stability of the polarization state of the probe.

The next step is metal coating and aperture formation. In general, the evaporated metal film generally has a grainy texture, resulting in an irregularly shaped aperture with nonisotropic polarization behavior. The grains also increase the distance between the aperture and the sample, not only degrading resolution but also reducing the intensity of the local excitation. As a method for making a high-definition aperture probe, we use a simple method based on the mechanical impact of the metal (Au) coated tip on a suitable surface [16, 17]. The resulting probe has a flat end and a well-defined circular aperture. Furthermore, the impact method assures that the aperture plane is strictly parallel to the sample surface, which is important in minimizing the distance between the aperture and the sample surface. The size of the aperture can be selected by carefully monitoring the intensity of light transmitting from the apex, since the throughput of the probe is strictly dependent on the aperture diameter.

1.3 Spatial Resolution of NSOM Studied by Single Molecule Imaging

Ultimate spatial resolution of NSOM is of great interest from the viewpoint of revealing the nature of light–matter interaction at the nanoscale. As a standard method for the evaluation of spatial resolution of NSOM, fluorescence imaging of a single molecule is most reliable because it behaves as an ideal point-like light source. Many groups have made efforts to improve the resolution in the single-molecule imaging using a variety of methods, such as apertureless NSOM [18] and a single molecule light source [19]. A spatial resolution as high as 32 nm has been reported in fluorescence imaging by using a microfabricated cantilevered probe [20]. By using an aperture probe, a spatial resolution as high as 25 nm has been reported recently in single molecule fluorescence imaging by scanning near-field optical/atomic force microscopy [21].

In this section, we describe single-molecule imaging with a high resolution of approximately 10 nm achieved by an aperture NSOM [22]. To discuss the depen-

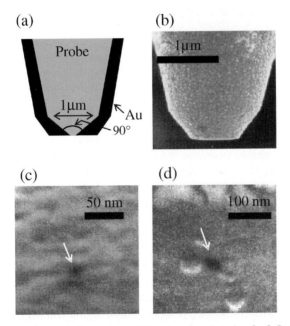

Fig. 1.3. a A cross-sectional illustration of the Au-coated probe. **b–d** Scanning electron micrographs of aperture probes: **b** a side view of a probe with the double-tapered structure; **c–d** apertures created by the impact method. Aperture diameters are 10 and 30 nm, respectively

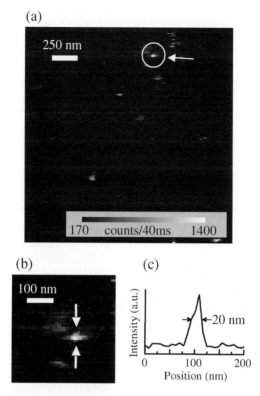

Fig. 1.4. a A fluorescence image of single Cy5.5 molecules at 633-nm excitation. **b** A magnified image of the bright spot circled in **a**. **c** A cross-sectional profile of the signal intensity evaluated along a line indicated by the pair of arrows in **b**. The spatial resolution determined from the FWHM of the profile is 20 nm

dence of the resolution on the wavelength of excitation light, measurements with two different excitation lasers for the same probes are carried out. These results are compared with a computational calculation employing the finite-difference time-domain (FDTD) method, which is appropriate for simulating electromagnetic field distributions applied to actual three-dimensional problems [23]. Thus we discuss the achievable spatial resolution of the aperture NSOM.

1.3.1 Single-Molecule Imaging with Aperture Probes

An NSOM fiber probe with the double-tapered structure and well-defined aperture created by the mechanical impact method, as described in Sect. 1.2, was employed. Samples examined were single dye molecules of Cy5.5 and Rhodamine dispersed on quartz substrates. Single-molecule dispersion on the substrate was confirmed by observing one-step photobleaching of almost all of the molecules. The fluorescence NSOM was operated in the illumination mode. As excitation light sources, a He–Ne laser ($\lambda = 633$ nm) and a SHG YVO$_4$ laser ($\lambda = 532$ nm) were employed. The

(a)

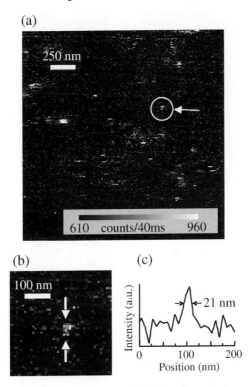

(b)   (c)

Fig. 1.5. a A fluorescence image of single Cy5.5 molecules at 532-nm excitation obtained using the same probe and the same sample as those in Fig. 1.4, but not measured in the same area as in Fig. 1.4(b). A magnified image of the bright spot circled in **a**. **c** A cross-sectional profile along a line indicated by the pair of arrows in **b**. The spatial resolution is estimated to be 21 nm

emission from a single dye molecule was collected by an objective lens and transported to an avalanche photodiode (APD) through a bandpass filter (center wavelength $\lambda = 700$ nm, bandwidth $\Delta\lambda = 40$ nm for the Cy5.5 dye, $\lambda = 600$ nm, $\Delta\lambda = 40$ nm for the Rhodamine dye). The sample–probe distance was controlled by a shear–force feedback mechanism.

Figure 1.3(a) shows a cross-sectional illustration of the Au-coated probe. Scanning electron micrographs of aperture probes are shown in Figs. 1.3(b)–(d): a side view of a probe with the double tapered structure and overhead views of apertures. From the scanning electron micrographs, which were taken after several scanning measurements, we found the probes have flat end-faces with small round apertures. The diameters of the apertures in Figs. 1.3(c) and 1.3(d) are estimated to be 10 nm and 30 nm, respectively.

Figure 1.4(a) shows a fluorescence image of single Cy5.5 dye molecules irradiated by the He–Ne laser light. Each bright spot is attributed to the fluorescence from a single molecule. The bright spot circled in the image is magnified in Fig. 1.4(b).

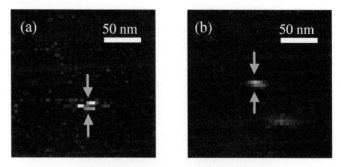

Fig. 1.6. The highest resolution images obtained with Rhodamine at 532-nm excitation (**a**) and Cy5.5 at 633-nm excitation (**b**). Estimated resolutions are 11 and 8 nm, respectively

Figure 1.4(c) shows a cross-sectional profile of the fluorescence signal intensity evaluated along a line indicated by a pair of arrows in Fig. 1.4(b). From the full width at half maximum (FWHM) of the profile, the spatial resolution is estimated to be 20 nm.

Figure 1.5(a) shows a fluorescence image obtained using the same probe and the same sample of single Cy5.5 dye molecules, but not measured in the same area as in Fig. 1.4(a), by the SHG YVO$_4$ laser excitation. The bright spot circled in Fig. 1.5(a) is magnified in Fig. 1.5(b), and Fig. 1.5(c) shows its cross-sectional profile. The spatial resolution estimated from the FWHM of the profiles is 21 nm.

The highest resolutions images obtained with Rhodamine at 532-nm excitation and Cy5.5 at 633-nm excitation are shown in Figs. 1.6(a) and 1.6(b), respectively. The resolution is estimated to be 11-nm at 532-nm excitation and 8-nm at 633-nm excitation.

1.3.2 Numerical Simulation of NSOM Resolution

To evaluate the achievable resolution of the aperture NSOM in visible range in the illumination mode of operation, a computer simulation by the FDTD method was employed for various aperture sizes and wavelengths. Electric fields (E) were calculated for the probe tip with an aperture diameter $D = 20$ nm at various wavelengths ($\lambda = 405, 442, 488, 514.5, 532$ and 633 nm) of irradiation lights, and for the probe tips with various aperture sizes ranging from $D = 0$ to 50 nm at the wavelength $\lambda = 633$ nm.

Figure 1.7(a) illustrates a cross-sectional view of the FDTD geometry of the three-dimensional problem, which reproduces the tip of the double-tapered probe with an aperture employed in the experiments. A three-dimensional illustration of the probe is shown in Fig. 1.7(b). The origin of the Cartesian coordinate was located at the center of the aperture. We assumed the light source, which was placed at 10 nm below the upper end of the tapered probe with a cone angle $\theta = 90°$, was a plane wave with a Gaussian distribution polarized along the x direction. The refractive index of the core of the fiber was 1.5 and the refractive indices of the real Au metal were extracted from [24]. The simulation box had a size of $1.6 \times 1.6 \times 0.8\,\mu m^3$ in

(a)

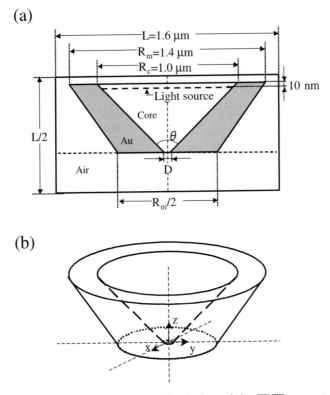

(b)

Fig. 1.7. a An illustration of the cross-sectional view of the FDTD geometry of three-dimensional problem, which reproduces the experimental situation. **b** A three-dimensional illustration of the tapered probe

the x, y and z directions. The space increment of the z directions around the aperture was 2 nm, and the increments of the x and y directions were 1 nm for aperture diameters less than $D = 10$ nm, and were 2 nm for the other aperture diameters.

Figures 1.8(a) and 1.6(b) show the intensity distribution of electronic field $|E|^2$ along the x- and y-axes, respectively, on $z = -4$ nm plane for the probe with the aperture of $D = 20$ nm at $\lambda = 633$-nm excitation. Here we define the spatial resolution of Δx and Δy as the FWHM of the intensity distribution, indicated by arrows in Figs. 1.8(a) and 1.8(b).

Spatial resolutions for the aperture of $D = 20$ nm at various wavelengths are plotted in Fig. 1.9. The skin depth of Au calculated from its optical constants is indicated by a dashed line. It is found that the dependence of the resolution on the excitation wavelength has a similar tendency as the skin depth of Au. Figure 1.10 shows the resolutions for various aperture diameters $D = 0$, 10, 20, 30 and 50 nm at $\lambda = 633$ nm. The result indicates that the discrepancy between the predicted resolution and the physical aperture size is less than 10 nm for $D > 10$ nm. The highest resolution is obtained at $D = 10$ nm and is evaluated to be $\Delta x = 16$ nm and $\Delta y = 12$ nm.

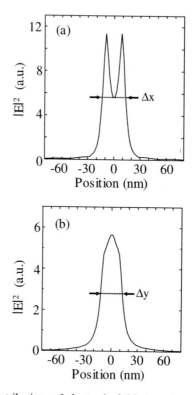

Fig. 1.8. The intensity distributions of electronic field along the x-axis **a** and the y-axis **b**, calculated on $z = -4$-nm plane for the probe with an aperture of $D = 20$ nm at $\lambda = 633$-nm excitation. The spatial resolutions of Δx and Δy are defined as the FWHM of the intensity distributions indicated by the pairs of arrows

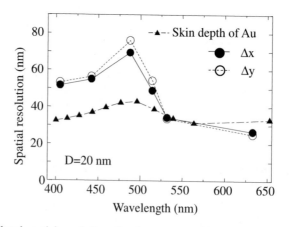

Fig. 1.9. Calculated spatial resolutions for the aperture with $D = 20$ nm at various wavelengths

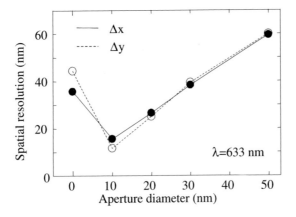

Fig. 1.10. Calculated spatial resolutions for various aperture diameters at $\lambda = 633$ nm

To discuss the resolution attainable using the NSOM with a tiny aperture, we compare the results of the computational calculation by the FDTD method with the experimental results obtained by the fluorescence imaging of single molecules. The same spatial resolutions as small as 20 nm were obtained experimentally at the different excitation wavelengths ($\lambda = 532$ and 633 nm) using the same aperture probe. The result does not agree with the results of the computational calculation for various excitation wavelengths in which about 10 nm of difference is predicted between the resolutions for the wavelengths of 532 and 633 nm. The dependence of the calculation results on the aperture sizes indicates that our computational simulation also does not reproduce the best resolutions in our measurements as high as 10 nm realized at excitations of both $\lambda = 532$ and 633 nm. The profile of intensity distribution of fluorescence signal obtained in the experimental operations is also greatly different from that evaluated along the x-axis in the computational calculation as characterized by the well-defined double peaks in Fig. 1.8. The disappearance of the double peaks can be explained by some distortion of the aperture shape. A slight inclination of the aperture face also results in contribution of a single peak because the intensity of the other peak decreases rapidly with the distance from aperture face. Taking account of the value of the FWHM of the intensity profile for one of the double peaks in Fig. 1.8, the experimental resolution as high as 10 nm is attributed to the efficient use of the localized near-field light with a single peak profile at the rim of the aperture.

1.4 Single Quantum Dot Spectroscopy and Imaging

In order to evaluate optical properties of QDs, such as an extremely sharp PL line, a macroscopic measurement, where an ensemble of QDs is observed at a time, is insufficient. This is because inhomogeneous broadening is inherent to QDs due to the distribution of their sizes and shapes. Thus, the intrinsic natures are hidden in

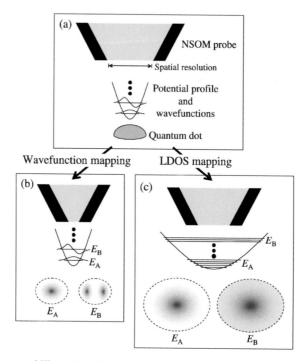

Fig. 1.11. Conceptual illustration of wavefunction mapping and LDOS mapping in single QD spectroscopy

an inhomogeneously broadened signal and spectroscopy on a single QD is strongly required. An NSOM offers a high spatial resolution, typically 100–200 nm, which is comparable to a typical dot-to-dot separation, and allows us to optically address individual quantum dots as illustrated in Fig. 1.11(a).

As described earlier, the size of light spot created by the NSOM probe is usually larger than the size of QD. Recent progress in the fabrication of aperture near-field fiber probe has pushed the spatial resolution to less than 30 nm [22, 25], which is comparable to or below the sizes of QDs. In such a case, NSOM allows us to investigate the inside of the QD. Roughly speaking, if the size of QD is smaller than 100 nm and the energy separation of discrete quantum levels is greater than the thermal energy at a cryogenic temperature, NSOM can visualize the spatial profile of a single quantum state: real-space mapping of an electron wavefunction [26–28]. This situation is illustrated in Fig. 1.11(b). Moreover, illumination of QD with an extremely narrow light source makes it possible to excite optically "dark" states whose excitation is forbidden by symmetry in the far field (breakdown of the usual optical selection rules) [10, 29]. These interesting observations and manipulation of electronic states in quantum confinement systems are unique to light–matter interaction at the nanoscale and the essential motivation for using the near-field optical method.

In another case we deal with a QD created by means of a nanofabrication technique. In contrast to naturally grown QDs, the size of artificially fabricated QDs can be as large as several hundreds of nm. In such a weakly localized electron systems, where energy separation of quantized states is smaller than thermal energy, NSOM maps out the local density of states as shown in Fig. 1.11(c). Spatially and energetically resolved spectroscopy is a powerful tool to reveal the localized and delocalized electron systems and, more importantly, their crossover region (weakly localized system).

1.5 NSOM Spectroscopy of Single Quantum Dots

1.5.1 Type II Quantum Dot

A self-assembled quantum dot is an ideal system for studying zero-dimensional quantum effects and has the potential for realizing future quantum devices. In self-assembled In(Ga)As/GaAs QDs with a band alignment classified as type I, both electrons and holes are confined in the QD. In a staggered type II band structure, the lowest energy states for an electron and a hole are concentrated on different layers [30–34]. Spatial separation occurs between the electron wavefunction in the GaAs layer and the hole wavefunction in a type II GaSb QD, and the optical properties differ from those of a type I QD [30, 32].

Single QD PL spectroscopy allows us to study multiexciton states by creating many excitons in a QD under high excitation conditions [35]. The two-exciton state is an especially interesting system, because it easily forms a bound biexciton state due to the attractive Coulomb interaction in a type IIn(Ga)As QD [36, 37]. The energy level of a bound biexciton state is lowered by the binding energy from the two excitons, where the binding energy is defined as the downward shift in energy of the biexciton relative to that of two uncorrelated excitons. As the stability of the biexciton state is sensitive to the structural and electronic parameters [38], the interaction between excitons in a type II GaSb QD should be different from that in a type IIn(Ga)As QD. Here we describe an experimental study of the exciton and two-exciton states in a single type II GaSb QD using the NSOM.

1.5.2 NSOM Spectroscopy of Single GaSb QDs

The sample in this study was self-assembled GaSb QDs grown on a GaAs (100) substrate using molecular beam epitaxy [39]. The lateral size, height and density of the GaSb QDs of an uncovered sample were 16–26 nm, 5–8 nm and 2×10^{10} cm^{-2}, respectively, as measured by an atomic force microscope. Cross-sectional transmission electron microscopy showed that a GaSb QD has a lens shape after capping with a GaAs cover layer of 100 nm. The sample was illuminated through the aperture with a diode laser (=685 nm) and the PL from a GaSb QD was collected via the same aperture. The PL signal was detected using a 32-cm monochromator equipped with a cooled charge-coupled device with a spectral resolution of 250 μeV. All measurements were conducted at cryogenic temperature.

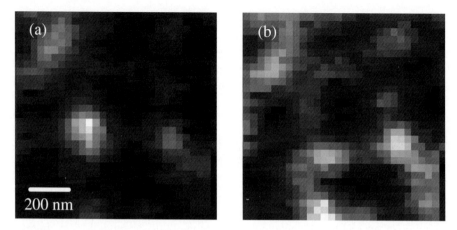

Fig. 1.12. Near-field PL images of single GaSb QDs monitored at photon energies of **a** 1.266 and **b** 1.259 eV, respectively

Figures 1.12(a) and 1.12(b) show typical near-field PL images, monitored at photon energies of 1.266 and 1.259 eV, respectively, under relatively low excitation conditions. Several bright spots of the PL signals from single GaSb QDs are observed in both images. We can confirm the spectroscopic observation of a single GaSb QD from the PL images. The average size of the bright spots, defined by the full width at half maximum (FWHM) of the PL intensity profile, is estimated to be about 120 nm, a value that corresponds to the spatial resolution of the measurement. The spatial resolution is somewhat larger than the aperture diameter of the probe tip (80 nm), because the GaSb QDs are embedded at a depth of 100 nm from the sample surface.

Figure 1.13 shows typical near-field PL spectra of the exciton emission from three different single GaSb QDs on an expanded energy scale. The linewidths of the three emission peaks are estimated to be 250 μeV, where the value is limited by the spectral resolution of the measurements. Consequently, the homogeneous linewidth of an exciton state in a type II GaSb QD is evaluated to be less than 250 μeV, which is narrower than the 280 μeV theoretically predicted in an ideal quantum well (QW) at 8 K. The narrow PL linewidth means that the exciton state in a type II GaSb QD has a longer coherence time than that in the QW.

Figure 1.14(a) shows near-field PL spectra of a single GaSb QD at various excitation power densities. A single emission peak in the PL spectrum, denoted as X, is observed at 1.2716 eV under lower excitation conditions (less than 1 μW). As shown in Fig. 1.14(b), the PL intensities of the X line, as a function of excitation power densities, show an almost linear power dependence under lower excitation conditions. The sharp, less than 1 meV FWHM linewidth, X emission line is assigned to the radiative recombination of the exciton consisting of a hole confined in a GaSb QD and an electron in the surrounding GaAs barrier layer, which are weakly bound together by an attractive Coulomb interaction.

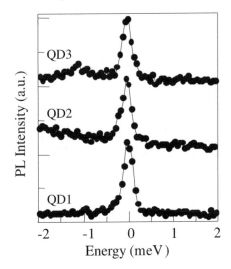

Fig. 1.13. PL spectra of three different single GaSb QDs at 8 K

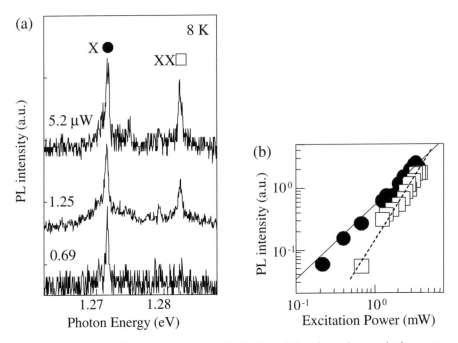

Fig. 1.14. a Near-field PL spectra of a single GaSb QD at 8 K under various excitation power densities. The PL peaks at 1.2716 and 1.2824 eV are denoted as X and XX, respectively. **b** Excitation power dependence of PL intensities of the X and XX lines. The *solid* (*dotted*) *line* corresponds to the gradient associated with linear (quadratic) power dependence

Next, in Fig. 1.14(a), we focus on the PL spectra at higher excitation conditions (greater than 1 μW). An additional peak appears at 1.2824 eV in the PL spectra, and the peak denoted as XX is observed at about 11 meV higher energy than the exciton emission (X). Figure 1.14(b) shows a nearly quadratic power dependence of the XX line as a function of the excitation power. The power dependence of the PL intensity suggests that the XX emission results from the radiative transition from a two-exciton state to the exciton ground state. In type I self-assembled In(Ga)As QDs [36] and naturally occurring GaAs QDs [37], the PL line is usually observed at 3–5 meV to the lower energy side of the exciton emission with the quadratic power dependence generally assigned to the bound biexciton emission. This experimental result, with the two-exciton emission occurring on the higher energy side of the exciton emission, contrasts the results in type I QDs. This is consistent with the results of the macroscopic PL spectra from GaSb QD ensembles showing a blueshift of the PL peak with increasing excitation power [30, 33].

The energy difference between the two-exciton emission (XX) and the exciton emission (X) corresponds to the binding energy ($E_{\text{bin}} = 2E_{\text{X}} - E_{\text{XX}}$), where E_{XX} and E_{X} are the energy of the two-exciton state and the exciton ground state, respectively. After measuring many GaSb QDs in the same sample, we found that E_{bin} always has negative values, ranging from -11 to -21 meV. A negative E_{bin} implies that the sign of the exciton–exciton interaction is repulsive in these QDs. In type II GaSb QDs, only the holes are confined inside the QD, while the electron wave function is relatively delocalized in the GaAs barrier layer around the QDs. Consequently, it is reasonable to expect the Coulomb energy of the two-exciton ground state to be mainly dominated by the hole–hole repulsive Coulomb interaction, and to have a negative value, because the strengths of the electron–hole and electron–electron interactions are smaller than that of the hole–hole interaction.

For a quantitatively accurate understanding, we performed theoretical calculations of two-exciton states in these QDs. We used the empirical pseudopotential model (EPM) that has been applied to various III–V type I QDs [40]. The single particle states are obtained by solving the one-electron Schrödinger equation in a potential, which is obtained from the superposition of atomic pseudopotentials centered at the location of each atom in a supercell containing the QD and the surrounding matrix. Spin–orbit coupling is included as a similar sum of nonlocal potentials [40]. The EPM parameters fitted to the bulk band structure parameters of GaSb and GaAs were taken from [41].

As described earlier, the exciton and two-exciton states involve electrons weakly bound to the QD solely by the Coulomb attraction of the confined holes. This situation makes it practically impossible to calculate the exciton and two-exciton states using the conventional configuration interaction approach typically used in type I QD calculations. To handle this situation, we developed a self-consistent mean field (SCF) calculation method for multiple electron–hole pair excitations within the EPM framework. In this approach, each single particle state of a multiexciton complex is calculated by including the Coulomb potential due to all the other particles occupying the lowest possible single particle orbitals. We use Resta's model [42] for the nonlocal dielectric constant. Our approach treats the electron–electron and electron–

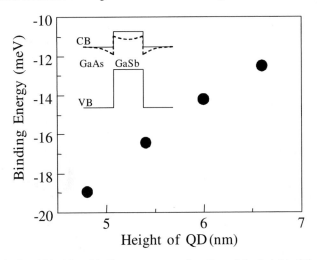

Fig. 1.15. Calculated biexciton binding energy as a function of the height of the lens shaped GaSb QDs. The height-to-diameter ratio was fixed as 0.3. The inset shows a schematic of the conduction band (CB) and valence band (VB) lineup of the GaSb/GaAs QDs. The dashed lines schematically illustrate the potential sensed by an electron when a hole is present

hole interactions at the Hartree–Fock level for one-exciton and two-exciton ground states and is identical to Hartree with a self-interaction correction for three or more exciton complexes.

First, we calculated the single particle energies and orbitals for a few lowest conduction and highest valence band states with zero, one and two electron–hole pairs using a linear combination of bulk Bloch functions as the basis [43]. The single- and two-exciton calculations are iterated to self-consistency. The exciton and two-exciton (biexciton) energies are calculated as the sum of single-particle energies corrected for double counting of the Coulomb interaction. A negative E_{bin} indicates that the two-exciton emission in PL spectra appears on the higher energy side of the exciton emission.

Although the structure is grown as nominally pure GaSb QDs in GaAs, independent studies have shown that relatively strong admixing of Sb and As atoms is expected [43]. Calculations were done for the lens-shaped $GaSb_{1-x}As_x$ QDs in a GaAs matrix. The absolute exciton energies depend strongly on the alloying, as well as the size and shape of the QDs. In addition, PL studies of GaSb/GaAs type II heterostructures tell us that the observed emission energies can be explained only by using a much smaller valence offset than is theoretically accepted [44]. Therefore, it is difficult to correlate the absolute exciton energies with the experiment. The calculated E_{bin} as a function of QD size is shown in Fig. 1.15. The calculated data correspond to QDs of heights ranging from 4.8–6.6 nm with the height-to-diameter ratio fixed at 0.3. We found E_{bin} from -12 to -19 meV, i.e., negative values for the entire range of QD sizes considered. The range of experimentally observed binding energies is very consistent with the calculated results. A detailed analysis of the results shows that although the two-exciton energy shift relative to the exciton could be understood qualitatively as due to the repulsion between the two confined holes, the contributions from electron–hole attraction and electron–electron repulsion are not negligible. For example, for a 4.8-nm high QD, the E_{bin} of -19 meV includes -27 meV of hole–hole repulsion, -5 meV of electron–electron repulsion, and $+12$ meV of electron–hole attraction.

1.6 Real-Space Mapping of Electron Wavefunction

With the recent progress in the nanostructuring of semiconductor materials and in the applications of these nanostructured materials in optoelectronics, NSOM microscopy and spectroscopy have become important tools for determining the local optical properties of these structures. In single quantum constituent spectroscopy, NSOM provides access to individual quantum constituent, such as QD, an ensemble of which exhibits inhomogeneous broadening due to the distribution of sizes, shapes and strains. NSOM can thus elucidate the nature of QD, including the narrow optical transition arising from the atom-like discrete density of states.

Single QD spectroscopy has revealed their long coherence times at low temperature and large oscillator strengths of optical transition. However, to improve these parameters for implementation of quantum computers, accurate information on the wavefunction for individual QDs is of great importance. In addition, in the study of coupled-QDs systems as interacting qubits, in which it is difficult to predict the exact wavefunction within theoretical frameworks, an optical spectroscopic technique for probing the wavefunction itself should be developed. By enhancing the spatial resolution of NSOM up to 10–30 nm, which is smaller than the typical size of QDs, local probing allows direct mapping of the real space distribution of the quantum eigenstate (wavefunction) within a QD, as predicted by theoretical studies [26–28].

In contrast to the well-defined quantum confined systems like QDs, the more common disordered systems with local potential fluctuations still leave open questions. To fully understand such complicated systems, exciton wavefunctions should be visualized with an extremely high resolution less than the spatial extension of wavefunction. NSOM, with a spatial resolution of 10 nm, is the only tool to obtain such information.

1.6.1 Light–Matter Interaction at the Nanoscale

In this section we summarize a theoretical approach to understand the light–matter interaction at the nanoscale [45]. When the nanoscale confined electron system, such as a semiconductor QD, is excited by light with a frequency ω, the absorbed power $\alpha(\omega)$ is

$$\alpha(\omega) \propto \int_E (r)P(r, \omega)\, dr, \tag{1.1}$$

where $E(r)$ is the spatial distribution of electromagnetic field and $P(r, \omega)$ is the induced interband polarization. In the general form the relationship between $P(r, \omega)$

and $E(r)$ should be expressed by the nonlocal electrical susceptibility $\chi(r, r'; \omega)$ as

$$P(r, \omega) = \int \chi(r, r'; \omega) E(r') \, dr', \tag{1.2}$$

$\chi(r, r'; \omega)$ can be obtained by eigenfunction ψ_{ex} and eigenenergy E_{ex} of exciton state confined in a QD:

$$\chi(r, r'; \omega) \propto \frac{\psi_{ex}(r) \psi_{ex}^*(r')}{E - \hbar\omega - i\gamma}. \tag{1.3}$$

Here we assume that quasi-resonant excitation at E_{ex} and therefore the contribution of other quantized exciton states are negligible. γ is a damping constant due to phonon scattering and radiative decay of exciton. By using (1.2) and (1.3), $\alpha(\omega)$ can be written in the form

$$\alpha(\omega) \propto \frac{|\int \psi_{ex}(r) E(r) dr|^2}{E - \hbar\omega - i\gamma}. \tag{1.4}$$

To illustrate the physical meaning in (1.4), we discuss two limiting cases. For far-field excitation, where QD is illuminated by a spatially homogeneous electromagnetic field, $\alpha(\omega)$ is given by the spatial integration of the exciton wavefunction,

$$\alpha(\omega) \propto \left| \int \psi_{ex}(r) \, dr \right|^2. \tag{1.5}$$

From the value of this integral, so-called optical selection rules are derived. If the integral is zero, the corresponding transition is "forbidden" and the exciton state is optically "dark". In the opposite limit of extremely confined light, $E(r)$ is assumed to be $\delta(r - R)$, where R is the position of the nanoscale light source, say a near-field tip. As a result one can probe the local value of the exciton wavefunction,

$$\alpha(\omega) \propto |\psi_{ex}(R)|^2. \tag{1.6}$$

By measuring $\alpha(\omega)$ as a function of the tip position, we can map out the exaction wavefunction. More interestingly, the dark-state exciton becomes visible by breaking the selection rule of optical transition. In the intermediate regime in terms of the confinement of light, ψ_{ex} are averaged over an illumination region.

Now we try to give an intuitive explanation on the local optical excitation using a classical coupled oscillator model as shown in Fig. 1.16. Each pendulum represents a localized dipole, such as a constituent molecule that makes up a molecular crystal. The dipole–dipole interaction, which forms an exciton as a collective excitation of constituent molecules, is taken into account by introducing springs to couple neighboring pendulums. Here we assume the size of the system (the size of molecular crystal) is much smaller than the wavelength of light. Figures 1.16(a) and 1.16(b) illustrate the lowest and the second lowest normal modes of the coupled oscillator, respectively. For the far-field illumination, all the pendulums are swung together at the frequency of irradiated light with the same phases. Therefore the second lowest mode, where two halves of pendulums move opposite, cannot be excited by the far-

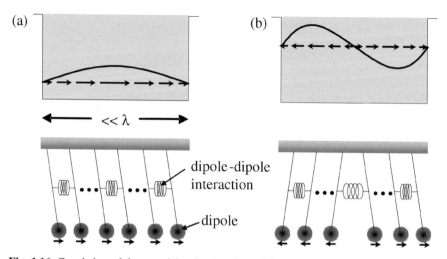

Fig. 1.16. Coupled pendulum model to intuitively explain the light–matter interaction at the nanoscale. **a** The lowest mode and **b** the second lowest mode of the coupled oscillator

field light whereas the lowest mode can be. This corresponds to the optical selection rule for far-field excitation of confined exciton systems. For the near-field regime, on the other hand, the situation drastically changes. The trick that the nanoscale confined light plays is to grasp solely a single pendulum and swing it. In this case, if the light frequency matches eigenfrequency of the individual oscillation mode, any normal mode can be excited regardless of the symmetry of oscillation, which means that the optical selection rule is broken by the near-field excitation. The efficiency of mode excitation is dependent on which pendulum is swung, i.e., the position of nanoscale light source. By swinging a pendulum in order from the end and observing the magnitude of mode oscillation for each we can map out the distribution of oscillation amplitude of individual pendulums. This illustrates the principle of the wavefunction mapping of exciton states.

1.6.2 Interface Fluctuation QD

Here we describe PL imaging spectroscopy of a GaAs QD by NSOM with a spatial resolution of 30 nm. This unprecedented high spatial resolution relative to the size of the QD (100 nm) permits a real-space mapping of the center-of-mass wavefunction of an exciton confined in the QD based on the principle discussed earlier [11, 46].

A schematic of QD sample structure is shown in Fig. 1.17. We prepared a 5-nm thick GaAs QW, sandwiched between layers of Al(Ga)As grown by molecular-beam epitaxy. Two-minute interruptions of the growth process at both interfaces lead to the formation of large monolayer-high islands which localize excitons in QD-like potential with lateral dimensions on the order of 40–100 nm [4]. The GaAs QW layer was covered with a thin barrier and a cap layer of totally 20 nm, allows the near-field tip to be close enough to the emission source (QD).

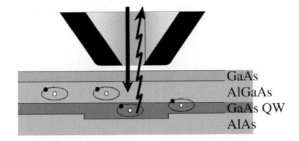

Fig. 1.17. A schematic of a GaAs quantum dot naturally formed in a quantum well due to the fluctuation of well thickness

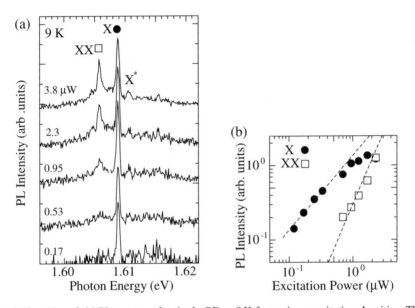

Fig. 1.18. a Near-field PL spectra of a single QD at 9 K for various excitation densities. The PL peaks at 1.6088, 1.6057 and 1.6104 eV are denoted by X, XX and X*. **b** Excitation power dependence of PL intensities of the X and the XX lines. The two dotted lines corresponds to the gradient associated with linear and quadratic power dependence

The GaAs QD was excited with He–Ne laser light ($\lambda = 633$ nm) through the aperture and carriers (excitons) were created in the barrier layers as well as the QW layer. Excitons diffused over several hundreds of nm and relaxed into the QDs. The PL signal from the QD was collected via the same aperture to prevent a reduction of the spatial resolution due to carrier diffusion. Near-field PL spectra were measured, for example, at 11 nm steps across a 210 nm × 210 nm area and two-dimensional images were constructed from a series of PL spectra.

Figure 1.18(a) shows near-field PL spectra of a single QD at 9 K at excitation densities ranging from 0.17 to 3.8 µW. At low excitation densities, a single emission line (denoted by X) at 1.6088 eV is observed. With an increase in excitation density,

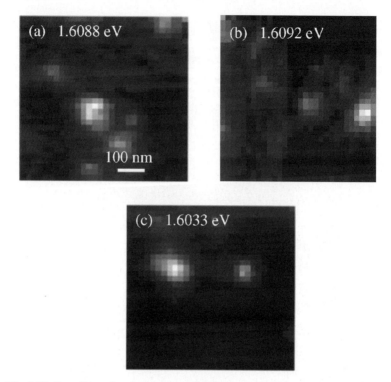

Fig. 1.19. Two-dimensional mapping of the PL intensity for three different X lines

the other emission lines appear at 1.6057 eV (XX) and at 1.6104 eV (X*). In order to clarify the origin of these emissions, we examined excitation power dependence of PL intensities as shown in Fig. 1.18(b). The X line can be identified as an emission from a single-exciton state by its linear increase in emission intensity and its saturation behavior. The quadratic dependence of the XX emission with excitation power indicates that XX is an emission from a biexciton state. This identification of the XX line is also supported by the difference in the emission energy of 3.1 meV, which corresponds to the binding energy of biexciton and agrees well with the values reported previously [47]. The X* emission line can be attributed to the radiative recombination of the exciton excited state by considering its energy position (higher energy side of the single exciton emission by about 1.6 meV) [48]. Figure 1.19 shows low-magnification PL maps for the intensity of X emissions with three different energies in the same scanning area. These emission profiles were found to differ from QD to QD.

1.6.3 Real-Space Mapping of Exciton Wavefunction Confined in a QD

The high-magnification PL images in Fig. 1.20 were obtained by mapping the PL intensity with respect to the X ((a), (c) and (e)) and the XX ((b), (d) and (f)) lines of

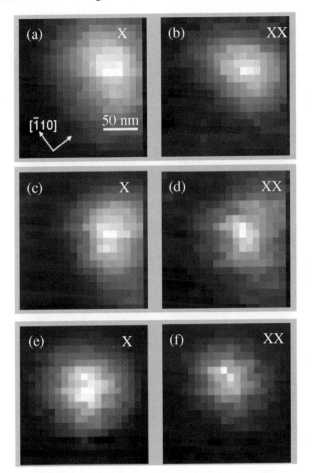

Fig. 1.20. Series of high-resolution PL images of exciton (X) state (**a, c** and **e**) and biexciton (XX) state (**b, d** and **f**) for three different QDs. Crystal axes along [110] and [−110] directions are indicated

three different QDs. The exciton PL images in Fig. 1.20 ((a), (c) and (e)) show an elongation along the [−110] crystal axis. The image sizes are larger than the PL collection spot diameter, i.e., the spatial resolution of the NSOM. The elongation along the [−110] axis due to the anisotropy of the monolayer-high island is consistent with previous observations with a scanning tunneling microscope (STM) [4]. We also obtained elongated biexciton PL images along the [−110] crystal axis in Fig. 1.20 ((b), (d) and (f)) and found a clear difference in the spatial distribution between the exciton and biexciton emission. Here the significant point is that the PL image sizes of biexcitons are always smaller than those of excitons.

Figures 1.21(a) and 1.21(b) show the normalized cross-sectional PL intensity profiles of exciton (thick lines) and biexciton (thin lines) along the [110] and [−110]

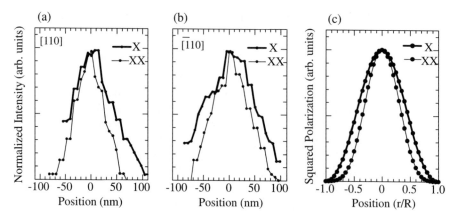

Fig. 1.21. High-resolution PL images and corresponding cross-sectional intensity profiles of the exciton (**a** and **b**) and the exciton excited (**c** and **d**) state. The intensity profiles are taken along *solid* and *dotted lines* in the images

crystal axes. The spreads in the exciton (biexciton) images, defined as the full width at half maximum (FWHM) of each profile are 80 (60) nm and 115 (80) nm along the [110] and [−110] crystal axes, respectively.

Theoretical considerations can clarify what we see in the exciton and biexciton PL images. The relevant quantity is the optical near-field around a single QD associated with an optical transition. This field can be calculated with Maxwell's equations using the polarization field of the exciton or biexciton as the source term. The observed luminescence intensity is proportional to the square of the near-field detected by the probe. In the following, however, we have calculated the emission patterns simply by the squared polarization fields without taking account of the instrumental details. The polarization fields at the position of the probe (r_s) are derived from the transition matrix element from the exciton state (X) to the ground state (0) and that from the biexciton state (XX) to the exciton state (X) as follows [49]:

$$\langle 0|p\delta(r - r_s)|X\rangle = -\sqrt{2}p_{cv}\phi(r_s, r_s), \tag{1.7}$$

$$\langle X|p\delta(r - r_s)|XX\rangle = -\sqrt{\frac{3}{2}}p_{cv}\sum_{r_1, r_a}\phi(r_1, r_a)\Phi^{++}(r_\Delta, r_s, r_a, r_s)$$

$$-\sqrt{\frac{1}{6}}p_{cv}\sum_{r_1, r_a}\phi(r_1, r_a)\Phi^{--}(r_\Delta, r_s, r_a, r_s), \tag{1.8}$$

where $\phi(r_e, r_h)$ stands for the exciton envelope function with the electron and hole coordinates denoted by r_e and r_h, $\Phi^{++}(\Phi^{--})(r_1, r_2, r_a, r_b)$ represents the biexciton envelope function with electron coordinates (r_1, r_2) and hole coordinates (r_a, r_b) that is symmetrized (antisymmetrized) with respect to the interchange between two electrons and between two holes, and p_{cv} is the transition dipole moment between the conduction band and the valence band. The spatial distribution of the exciton polarization field corresponds to the center-of-mass envelope function of a confined exciton. For the biexciton emission, the polarization field is determined by the overlap integral, which represents the spatial correlation between two excitons forming the biexciton and is expected to be more localized than the single exciton wavefunction.

Figure 1.21(c) shows the squared polarization amplitudes of the exciton (thick line) and biexciton (thin line) emission, which have been calculated for a GaAs QD with size parameters relevant to our experiments. The calculated profile of the squared polarization amplitude of the biexciton emission is narrower than that of the exciton emission. The spread of the biexciton emission normalized by that of the exciton emission is estimated to be 0.76, which is in good agreement with the experimental result (0.75 ± 0.08). This theoretical support and the experimental facts lead to the conclusion that the local optical probing by the near-field scanning optical microscope directly maps out the center-of-mass wavefunction of an exciton confined in a monolayer-high island.

Furthermore, we can demonstrate a novel powerful feature of the wavefunction mapping spectroscopy. Figure 1.22 shows the PL image and corresponding cross-sectional intensity profiles of the exciton ground state X ((a) and (b)) and the exciton excited state X* ((c) and (d)) from a single QD, which is different from that observed in Fig. 1.21. The exciton PL image exhibits a complicated shape in this QD, unlike the simple elliptical shape shown in Fig. 1.21. This is because the exciton is confined in a monolayer-high island with an extremely anisotropic shape. The significant point is that the exciton ground state image exhibits a single maximum peak in the intensity profile, while a double-peaked intensity profile is obtained from the exciton excited state. This is attributable to the difference in spatial distribution of the center-of-mass wavefunction, which has no node in the ground state, but does have a node in the excited state.

1.7 Real-Space Mapping of Local Density of States

Since the local electronic structure—defined as the local density of states (LDOS) in metal corrals—was first demonstrated using scanning probe microscopy and spectroscopy [50, 51], the LDOS mapping technique has been applied to many interesting quantum systems, such as two-dimensional (2D) electron gas [52], one-dimensional quantum wires [53] and zero-dimensional (0D) quantum dots (QDs) [54]. Although NSOM is useful for studying the elementary excitation of these quantum structures with less than subwavelength spatial resolution, there are only a few results of LDOS mapping using NSOM: for example, observation of the optical LDOS of an optical corral structure with a forbidden light [55].

Here we probed the local electronic states of a Be-doped GaAs/Al1-xGaxAs single heterojunction with a surface gate using an NSOM. The spatial distribution of LDOS in a field-induced quantum structure can be mapped using near-field PL microscopy, as the quantum structure investigated here is larger than the spatial resolution of NSOM and the PL spectrum reflects the DOS of electrons.

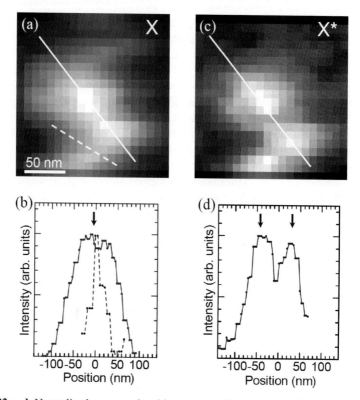

Fig. 1.22. a, b Normalized cross-sectional intensity profiles of exciton (thick lines) and biexciton (thin lines) PL images corresponding to Figs. 1.20(a) and (b). **c** Spatial distributions of squared polarization fields of the exciton (thick line) and biexciton (thin line) emission, which are theoretically calculated for a GaAs quantum dot (radius of 114 nm, thickness of 5 nm). The horizontal axis is normalized by the disk radius R

1.7.1 Field-Induced Quantum Dot

The QDs formed by an electrostatic field effect have been extensively studied [56–59]. In a field-induced QD, the strength and lateral profile of the confinement potential can be tuned using the design of the surface gate and the strength of the bias-voltage applied to the surface gate. As the degradation and imperfections at interfaces are minimized owing to electrostatic confinement, the electrons are confined by the well-defined lateral potential in this system. The properties of confined electrons have been investigated using macroscopic PL spectroscopy in a field-induced quantum structure based on a Be-doped single heterojunction [59, 60]. In this characteristics structure, the PL spectrum arising from the recombination of holes bound to Be acceptors with electrons in an electron gas provides us with a probe to investigate the DOS of electrons owing to relaxation of the k-selection rule in the optical process [59, 61, 62].

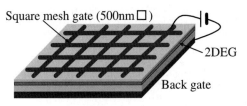

Fig. 1.23. A schematic of AlGaAs/GaAs two-dimensional electron gas with a mesh gate structure

The sample investigated in this study was a Be-doped single heterojunction of a GaAs/Al$_{1-x}$Ga$_x$As ($x = 0.7$) structure fabricated using molecular beam epitaxy [59, 62]. Figure 1.23 illustrates a rough schematic of sample structure. The heterostructure was grown on an n-type GaAs substrate used as the back contact and was fabricated under 75 nm from the surface. The nominal concentration of Be dopant was 2.0×10^{10} cm^{-2} and the Be-doped layer was inserted 25 nm below the heterojunction interface. The estimated sheet electron density without modulation using an external bias-voltage (V_B) was 3.6×10^{11} cm^{-2} at 1.8 K, using an optical Shubnikov–de Hass measurement. A semitransparent Ti/Au Schottky gate structure on the surface was fabricated with a square mesh of a 500-nm period using electron beam lithography. The bias-voltage was applied between the surface Schottky gate and the Ohmic back electrode.

An aperture about 120 nm in diameter was fabricated by milling of the probe apex using a focused ion beam (FIB) apparatus. The sample on the scanning stage was illuminated with a cw diode laser light (=685 nm) through the aperture, and a time-integrated PL signal from the sample was collected via the same aperture. The PL signal was sent to a 32-cm monochromator with a cooled charge coupled device with a spectral resolution of 220 μeV. The spatial resolution of NSOM in this study was about 140 nm.

Figure 1.24 shows far-field PL spectra, measured at the V_B ranging from 0 to -1.6 V. The PL signal from the 1.475 to 1.491 eV region is attributed to the recombination between the localized holes bound to Be acceptors with electrons in the electron gas. The holes bound to Be acceptors can recombine with any electrons with wave vectors up to the inverse of the hole Bohr radius with nearly equal optical transition probabilities [59–62]. Owing to the small effective Bohr radius of the hole bound to the Be acceptor, the PL spectrum of the 1.475–1.491 eV region reflecting the DOS of electrons [59] is used as a probe to investigate the electronic structure. Under low negative bias conditions ($V_B < -0.35$ V), the signals from 1.480 to 1.489 eV show flat shape PL spectra reflecting the 2D DOS. The strong peaks at 1.496 eV come from the recombination between 2D electrons in the second subband with holes bound to the acceptors [60]. When the V_B is increased to -1.6 V, the PL spectra show a linear increase in intensity toward higher photon energies from 1.480 to 1.489 eV. This behavior is expected from the 0D DOS of electrons because the linear dependence is in accordance with the generally accepted picture, in which the degeneracy of states increases with the quantum number.

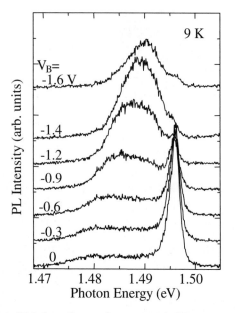

Fig. 1.24. Bias-voltage (V_B) dependence of macroscopic PL spectra of a Be-doped single heterojunction

1.7.2 Mapping of Local Density of States in a Field Induced QD

We investigated the local electronic states of the field-induced quantum structure while tuning the external bias voltage. As the PL intensity owing to the recombination of holes with electrons in an electron gas is proportional to the amplitude of the DOS and the spatial resolution of NSOM is higher than the size of the quantum structure, we can map the LDOS experimentally by monitoring the spatial distribution of the PL intensity from 1.475 to 1.491 eV. The atomic force microscopy image of a gated sample surface in Fig. 1.25(a) shows a square mesh gate with a 500-nm period. Figures 1.25(b)–2(e) show near-field optical images obtained by detecting the PL intensity at around 1.483 eV while changing the external bias-voltage (V_B = 0, −1.2, and −1.6 V). In a series of images, we can observe the change in the PL images from 2D (plane) to 0D (dot) features with the application of V_B to the surface gate. For V_B = −1.6 V, a bright spot is observed in the center of the square mesh gate in the PL image in Fig. 1.25(d). Looking at a wide spatial area, we see an array of PL spots corresponding to the period of the square mesh, as shown in Fig. 1.25(e). The change induced by applying a bias voltage is also supported by the cross-sectional intensity profiles shown in Fig. 1.25(f), taken along a diagonal of the mesh gate at the same positions. The size of the full width at half maximum (FWHM) of the profiles decreases from 400 nm at V_B = 0 V to 160 nm at V_B = −1.6 V. The narrow distribution of the PL intensity is caused by depletion of the electron density in the electron gas around the mesh gate under the external bias voltage. As a result, there is a dense electron population at positions far from the mesh gate and the potential

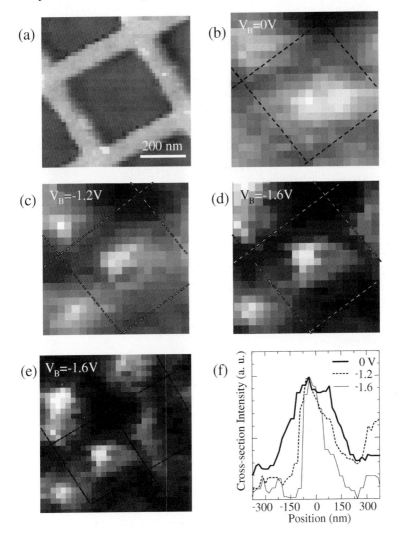

Fig. 1.25. a Shear-force microscopy (topographic) image of the gated sample surface (height contrast: 50 nm). **b–d** Near-field PL images at different bias voltages V_B = 0, −1.2, and −1.6 V, respectively. These images were monitored at a detection energy of around 1.483 eV at 9 K. The dotted lines in the images correspond to the position of the surface gate. **e** Near-field PL image at V_B = −1.6 V, measured for a wide area. **f** Cross-sectional PL intensity profiles taken along a diagonal of the mesh gate

for electrons is minimal at the center of the mesh gate. Therefore, the change in the PL image directly connects to the change in electronic structure from a 2D electron gas to the confined 0D electronic state (QD) and an artificially formed QD array, induced by the electrostatic confinement potential.

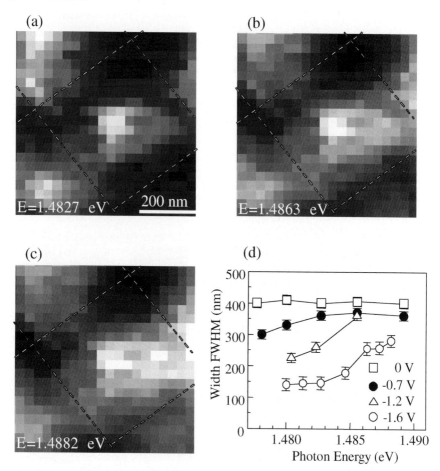

Fig. 1.26. a–c Near-field PL images obtained at different detection energies under a bias volt-age of −1.6 V. **d** PL intensity distribution defined as the FWHM of the profiles as a function of the detection photon energy under various bias-voltage conditions

Figures 1.26(a)–1.26(c) show PL images in the QD state under $V_B = -1.6$ V, detected at 1.4827, 1.4863 and 1.4882 eV, respectively. The spatial distribution of the PL intensity changes with the monitored photon energy and gradually spreads, going from an image at a lower photon energy to one at a higher photon energy [from Fig. 1.26(a)–1.26(c)]. We evaluated the spread of the PL image defined as the FWHM of the intensity profile as a function of photon energy and plotted the values for various bias voltages in Fig. 1.26(d). At low bias voltage (−0.7 V), the values of the spread in the PL images are essentially constant for the entire energy range from 1.477 to 1.490 eV, which is easily understood from the 2D DOS characteristics. By contrast, under higher bias voltage, the spread of the PL image strongly depends on the monitored energy and the value increases gradually toward the higher energy side, which indicates that the distribution of the LDOS gradually spreads from lower to higher energy states in a field-induced QD.

To confirm the feasibility of the LDOS mapping, we refer the numerical calculation results of the electron density distribution derived from solving Schrödinger and Poisson's equations [62–65]. The calculated potential for electrons in this quantum structure is minimal at the center of the mesh gate with an application of the bias voltage [62, 63]. The electrons with lower energy near the bottom of the electrostatic potential are confined strictly and the spatial distribution of the wave function extends with increasing energy. The experimental results obtained from the near-field PL images are consistent with the calculated electron density distributions and its energy dependence. Thus, the optical near-field microscopy maps the LDOS in a field-induced quantum structure.

Finally, we will mention the near-field PL spectrum of a field-induced single QD (not shown here). We did not observe the sharp spectral features, as frequently observed for 0D systems (QDs) [66, 67]. A peak in the PL spectrum arising from each confined level should be broadened by at least 0.5 meV, taking into consideration the estimated energy separation between confined levels [63]. This broadening might be because it takes $0.1–1\,\mu s$ for the nonequilibrium electrons to cool after excitation [61]. Therefore, the combination of near-field PL microscopy with the time-gated PL detection technique will enable us to observe fine spectral structures of a field-induced QD.

1.8 Carrier Localization in Cluster States in GaNAs

1.8.1 Dilute Nitride Semiconductors

In contrast to the well-defined quantum confined systems such as QDs grown in a self-assembled mode, the more common disordered systems with local potential fluctuations leave unanswered questions. For example, a large reduction of the fundamental band gap in GaAs with small amounts of nitrogen is relevant to the clustering behavior of nitrogen atoms and resultant potential fluctuations [68]. NSOM characterization with high spatial resolution can give us a lot of important information that is useful in our quest to fully understand such complicated systems, such as details about the localization and delocalization of carriers, which determine the optical properties in the vicinity of the band gap.

Dilute GaNAs and GaInNAs alloys are promising materials for optical communication devices [69–71] because they exhibit large band-gap bowing parameters. In particular, for long-wavelength semiconductor laser application, high temperature stability of the threshold current is realized in the GaInNAs/GaAs quantum well as compared to the conventional InGaAsP/InP quantum well due to strong electron confinement. However, GaNAs and GaInNAs with a high nitrogen concentration of more than 1% have been successfully grown only under nonequilibrium conditions by molecular beam epitaxy and metal organic vapor phase epitaxy.

The incorporation of nitrogen generally induces degradation of optical properties. To date, several groups of researchers have reported characteristic PL properties of GaNAs and GaInNAs, for example, the broad asymmetric PL spectra and the anomalous temperature dependence of the PL peak energy. More seriously emission yield drastically degraded with an increase of nitrogen concentration.

For improvement of their fundamental optical properties, it is strongly required to clarify electronic states due to single N impurities [72, 73] or N clusters, which interact with each other and with the host states. The interaction gives rise to the formation of weakly localized and delocalized electronic states at the band edge. Hence the shape of optical spectra is extremely sensitive to the N composition. In conventional spatially resolved PL spectroscopy, it is easy to detect single impurity emissions in ultradilute compositional region. However, in order to resolve complicated spectral structures and to clarify the interaction of localized states and the onset of alloy formation, a spatial resolution far beyond the diffraction limit is needed.

Here we show the results of spatially resolved PL spectroscopy with a high spatial resolution of 30 nm. Spatial inhomogeneity of PL is direct evidence of carrier localization in the potential minimum case caused by the compositional fluctuation. PL microscopy with such a high spatial precision enables the direct optical observation of compositional fluctuations, i.e., spontaneous N clusters and N random alloy regions, which are spatially separated in GaN_xAs_{1-x}/GaAs QWs.

1.8.2 Imaging Spectroscopy of Localized and Delocalized States

The samples investigated in this study were 5-nm thick GaN_xAs_{1-x}/GaAs single QWs with different N compositions ($x = 0.7\%$ and 1.2%) grown on (001) GaAs substrates using low-pressure metalorganic vapor phase epitaxy [74]. The growth temperature was $510\,°C$ and the details of the growth conditions have been described elsewhere [74]. The GaNAs layer was sandwiched between a 200-nm thick GaAs buffer layer and a 20-nm thick GaAs barrier layer. The thin 20-nm thick barrier layer allowed a near-field probe tip to come close enough to the emission sources to achieve a spatial resolution as high as 35 nm. The N composition (x) of the QW layer was estimated using secondary ion mass spectroscopy and cross-checked using the energy position in the PL spectra [75]. After growth, thermal annealing was performed for 10 min in a mixture of H_2 and TBAs at $670\,°C$ to improve the PL intensity [74].

We used NSOM probe tips with apertures of different diameters (30 and 150 nm), depending on the measurements. Optical measurements were performed at 8 K with a setup similar to that described earlier. Near-field PL spectra were obtained at every 12 nm steps for a 300 nm × 300 nm area, and two-dimensional PL maps were constructed from a series of these spectra.

The dotted line in Fig. 1.27 shows a far-field PL spectrum of a single GaN_xAs_{1-x}/GaAs ($x = 0.7\%$) QW at low temperature. The far-field spectrum has a broad linewidth of 30 meV and a lower-energy tail. To resolve the inhomogeneously broadened PL spectrum, we carried out near-field PL measurements with a high spatial resolution of 35 nm. The near-field PL spectrum in Fig. 1.27 has fine structures that

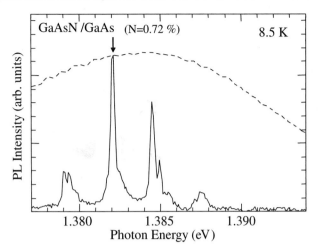

Fig. 1.27. Far-field (dotted line) and near-field (solid line) PL spectra of a GaN_xAs_{1-x}/GaAs ($x = 0.72\%$) single QW at 8.5 K

are not observed in the far-field spectrum [76–78]. After analyzing several thousands of near-field PL spectra, we found that the fine structures in the near-field PL spectra were divided into two groups: Sharp luminescence peaks with narrow linewidths below 1 meV and broad peaks with linewidths of several meV. We discuss the origin of these spectral features using both spectral and spatial information.

Figure 1.28(a) shows a typical near-field PL spectrum with sharp emission lines. To evaluate the linewidth, we show one of the sharp emission lines (1.382 eV) at an expanded energy scale in the inset; the spectral linewidth, defined as the full width at half maximum (FWHM), was determined to be less than 220 µeV, which is limited by the spectral resolution. The narrow PL linewidth means that the exciton state has a long coherence time, i.e., there is a reduction of the scattering rate between an exciton and phonons due to the change in the electronic structures from a continuum to discrete density of states. Such discrete density of states might be explained by the formation of naturally occurring quantum dot (QD) structures in a narrow GaN_xAs_{1-x}/GaAs QW ($x = 0.7\%$).

The spatial characteristics of the naturally occurring QD structures in a narrow QW showing the sharp emissions should be investigated. Figure 1.28(b) shows a high-resolution optical image of the sharp PL line, obtained by mapping the intensity (denoted by the arrow in Fig. 1.27). The surface topography did not influence the optical images, because the sample had a flat surface with a roughness of less than several nm, as estimated from a shear-force topographic image. The PL image clearly shows a point-emission feature, and the spot size defined as the FWHM of the cross-sectional profile shown in Fig. 1.28(c) is estimated to be 35 nm, which is limited by the spatial resolution of NSOM. The experimental spatial and spectral results suggest that the exciton strongly localizes in a potential minimum of naturally occurring QD structure in a narrow QW. The local N-rich regions (spontaneous N clusters) in a

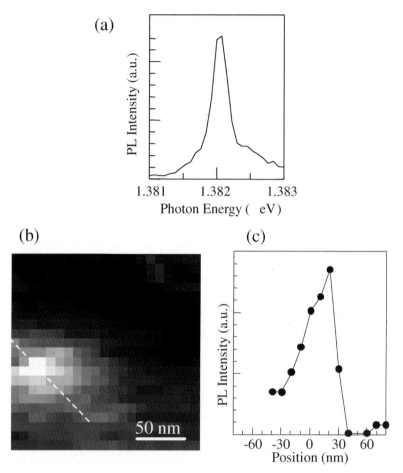

Fig. 1.28. a Near-field PL spectrum with a sharp emission peak. **b** Two-dimensional PL intensity mapping of the sharp emission line at 1.382 eV. **c** Cross-sectional PL intensity profile taken along the dotted line in **b**. The size of the emission profile, defined as the FWHM, is 35 nm (restricted by the spatial resolution of NSOM)

GaNAsN layer are the origin of the naturally occurring QD structures, as indicated by transmission electron microscopy [79].

Consider the broad peaks in the near-field PL spectrum. Figure 1.29(a) shows a typical PL spectrum of a broad peak with a 3 meV linewidth, which is much broader than below 220 μeV for the sharp emission line. The two-dimensional PL intensity map of the broad emission line in Fig. 1.29(b) shows spatially extended behavior that extends approximately 80 nm, as estimated from the cross-sectional profile in Fig. 1.29(c) (taken along the dotted line in the PL image). These characteristics of the broad PL line indicate that the exciton has a delocalized nature due to the random alloy state, which is frequently observed in isovalent semiconductor alloys. In addi-

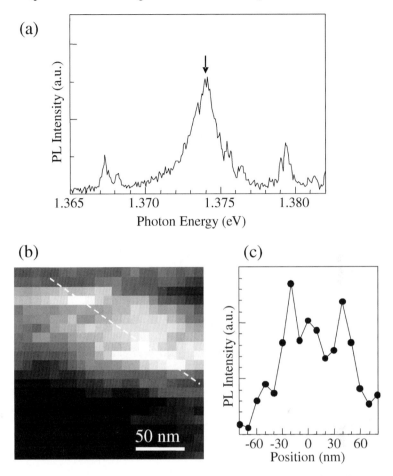

Fig. 1.29. a Near-field PL spectrum with a broad emission peak. **b** PL intensity map of the broad emission line at $1.374\,\text{eV}$. The scanning area is same as that of Fig. 1.1(b). **c** Cross-sectional PL intensity profile of the broad emission along the dotted line in **b**. The FWHM of the profile is $80\,\text{nm}$

tion, the energy positions and linewidths in the PL spectra are unchanged throughout the bright regions in Fig. 1.29(b), which supports the delocalized nature of excitons. Note that the PL images in Figs. 1.28(b) and 1.29(b) were obtained in the same scanning area. Therefore, the regions with randomly distributed N and the spontaneous N-rich clusters are separated spatially in $\text{GaN}_x\text{As}_{1-x}$ ($x = 0.7\%$), as observed directly using NSOM with a high spatial resolution of $35\,\text{nm}$.

We investigated the compositional fluctuations in $\text{GaN}_x\text{As}_{1-x}$ for different concentrations of N. Figure 1.30 shows near-field PL spectra for $x = 1.2\%$ at different spatial positions. These PL spectra were obtained using a probe tip with a 150-nm aperture, because the PL intensity strongly depends on x [77] and the intensity at

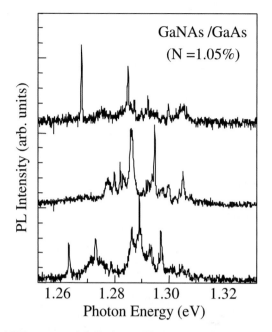

PL Intensity (arb. units)

GaNAs /GaAs
(N =1.05%)

1.26   1.28   1.30   1.32
Photon Energy (eV)

Fig. 1.30. Near-field PL spectra of GaN_xAs_{1-x}/GaAs ($x = 1.2\%$). The PL spectra measured at different positions were normalized by the maximum peak intensity

higher x ($=1.2\%$) decreases by about one order of magnitude compared with that for $x = 0.7\%$. The near-field PL spectra in Fig. 1.30 consist of many sharp lines from localized exciton recombinations in N-rich clusters and broad emissions, similar to the near-field PL spectra for $x = 0.7\%$. Note that sharp emission peaks are observed in the PL spectra for $x > 1\%$. Recent macroscopic PL [80] and magneto-PL [77] measurements of GaN_xAs_{1-x} for $x = 1.0\%$ showed broad, smooth spectra, which were assigned to emissions from delocalized excitons. However, our near-field PL spectroscopy reveals that the inhomogeneous broadened PL spectrum consists of the sharp emission lines owing to the recombinations of excitons localized in N-rich clusters superimposed on the broad emissions from the delocalized exciton state in GaN_xAs_{1-x} ($x = 1.2\%$). The regions of randomly distributed N and spontaneous N-rich clusters coexist at N compositions over 1%.

1.9 Perspectives

The dramatic progress in the spatial resolution of near-field optical microscopes offers an exciting opportunity for the study of light–matter interaction at the nanoscale. Real-space imaging of spatial distributions of quantized states can answer fundamental questions about the localized and delocalized nature of electrons in complicated potential systems, such as disorder alloy semiconductors. Ultimately, small light spots affect the light–matter interaction through the modification of quantum interference. This allows us to break the optical selection rule and to excite dark states.

A nanoscale light source also provides new techniques for wave packet engineering; creation, detection, transport, tailoring and coherent control of the electron wave packet. The pronounced quantum features of wave packet dynamics in nanoscale length will open new possibilities for controlling the capture processes from delocalized states to localized states. In combination with spin degree of freedom, control of spin-polarized wave packets leads to the realization of nano–spintronic devices.

Acknowledgments

We are grateful to M. Ohtsu, S. Mononobe, K. Matsuda, N. Hosaka, M. Sakai, K. Sawada, H. Nakamura, Y. Aoyagi, M. Mihara, S. Nomura, S. Nair, T. Tagahara, M. Takahashi, A. Moto, and S. Takagishi for their assistance and fruitful discussions.

References

[1] Ohtsu, M., Kobayashi, K., Kawazoe, T., Sangu, S., Yatsui, T.: IEEE J. Sel. Top. Quantum Electron. **8**, 839 (2002)

[2] P. Micheler, *Single Quantum Dots* (Springer, Berlin, 2003)

[3] D. Gammon, E.S. Snow, B.V. Shanabrook, D.S. Katzer, D. Park, Phys. Rev. Lett. **76**, 3005 (1996)

[4] J. Hours, P. Senellart, E. Peter, A. Cavanna, J. Bloch, Phys. Rev. B **71**, R161306 (2005)

[5] M. Ohtsu, *Near-Field Nano/Atom Optics and Spectroscopy* (Springer, Tokyo, 1998)

[6] L. Novotny, B. Hecht, *Principles of Nano-Optics* (Cambridge University Press, New York, 2006)

[7] N. Hosaka, T. Saiki, Opt. Rev. **13**, 262 (2006)

[8] J.N. Farahani, D.W. Pohl, H.-J. Eisler, B. Hecht, Phys. Rev. Lett. **95**, 17402 (2005)

[9] K. Cho, *Optical Response of Nanostructures* (Springer, Berlin, 2003)

[10] K. Matsuda, T. Saiki, S. Nomura, M. Mihara, Y. Aoyagi, S. Nair, T. Takagahara, Phys. Rev. Lett. **91**, 177401 (2003)

[11] K. Karrai, R.D. Grober, Appl. Phys. Lett. **66**, 1842 (1995)

[12] T. Saiki, K. Nishi, M. Ohtsu, Jpn. J. Appl. Phys. **37**, 1638 (1998)

[13] H. Nakamura, T. Sato, H. Kambe, K. Sawada, T. Saiki, J. Microscopy **202**, 50 (2001)

[14] T. Saiki, S. Mononobe, M. Ohtsu, N. Saito, J. Kusano, Appl. Phys. Lett. **68**, 2612 (1996)

[15] D.W. Pohl, W. Denk, M. Lanz, Appl. Phys. Lett. **44**, 651 (1984)

[16] T. Saiki, K. Matsuda, Appl. Phys. Lett. **74**, 2773 (1999)

[17] P. Anger, P. Bharadwaj, L. Novotny, Phys. Rev. Lett. **96**, 113002 (2006)

[18] J. Michaelis, C. Hettich, J. Mlynek, V. Sandoghdar, Nature **405**, 325 (2000)

[19] R. Eckert, J.M. Freyland, H. Gersen, H. Heinzelmann, G. Schurmann, W. Noell, U. Staufer, N.F. de Rooij, Appl. Phys. Lett. **77**, 3695 (2000)

[20] J.M. Kim, T. Ohtani, H. Muramatsu, Surf. Sci. **549**, 273 (2004)

[21] N. Hosaka, T. Saiki, J. Microscopy **202**, 362 (2001)

[22] H. Furukawa, S. Kawata, Opt. Commun. **132**, 170 (1996)

[23] E.D. Palik, *Handbook of Optical Constants of Solids* (Academic Press, New York, 1985)

[24] K. Matsuda, T. Saiki, S. Nomura, M. Mihara, Y. Aoyagi, Appl. Phys. Lett. **81**, 2291 (2002)

[25] G.W. Bryant, Appl. Phys. Lett. **72**, 768 (1998)

[26] Di Stefano, S. Savasta, G. Pistone, G. Martino, R. Girlanda, Phys. Rev. B **68**, 165329 (2003)

[27] E. Runge, C. Lienau, Appl. Phys. B **84**, 103 (2006)

[28] U. Hohenester, G. Goldone, E. Molinari, Phys. Rev. Lett. **95**, 216802 (2005)

[29] F. Hatami, M. Grundmann, N.N. Ledentsov, F. Heinrichsdorff, R. Heitz, J. Bohrer, D. Bimberg, S.S. Ruvimov, P. Werner, V.M. Ustinov, P.S. Kop'ev, Zh.I. Alferov, Phys. Rev. B **57**, 4635 (1998)

[30] K. Suzuki, R.A. Hogg, Y. Arakawa, J. Appl. Phys. **85**, 8349 (1999)

[31] Ph. Lelong, K. Suzuki, G. Bastard, H. Sakaki, Y. Arakawa, Physica E **7**, 393 (2000)

[32] L.M. Kircsch, R. Heitz, A. Schliwa, O. Stier, D. Bimberg, H. Kirmse, W. Neumann, Appl. Phys. Lett. **78**, 1418 (2001)

[33] E. Ribeiro, A.O. Govorov, W. Carvalho, G. Medeiros-Ribeiro, Phys. Rev. Lett. **92**, 126402 (2004)

[34] E. Dekel, D. Gershoni, E. Ehrenfreund, D. Spektor, J.M. Garcia, P.M. Petroff, Phys. Rev. Lett. **80**, 4991 (1998)

[35] D.V. Regelman, U. Mizrahi, D. Gershoni, E. Ehrenfreund, W.V. Schoenfeld, P.M. Petroff, Phys. Rev. Lett. **87**, 257401 (2001)

[36] K. Brunner, G. Abstreiter, G. Bohm, G. Trankle, G. Weimann, Phys. Rev. Lett. **73**, 1138 (1994)

[37] S. Rodt, R. Heitz, A. Schliwa, R.L. Sellin, F. Guffarth, D. Bimberg, Phys. Rev. B **68**, 035331 (2003)

[38] T. Nakai, S. Iwasaki, K. Yamaguchi, Jpn. J. Appl. Phys. **43**, 2122 (2004)

[39] A.J. Williamson, L.W. Wang, A. Zunger, Phys. Rev. B **62**, 12963 (2000)

[40] R. Resta, Phys. Rev. B **16**, 2717 (1977)

[41] L.-W. Wang, J. Kim, A. Zunger, Phys. Rev. B **59**, 5678 (1999)

[42] T. Nakai, K. Yamaguchi, Jpn. J. Appl. Phys. **44**, 3803 (2005)

[43] N.N. Ledentsov, J. Bohrer, M. Beer, F. Heinrichsdorff, M. Grundmann, D. Bimberg, S.V. Ivanov, B.Ya. Meltser, S.V. Shaposhnikov, I.N. Yassievich, N.N. Faleev, P.S. Kop'ev, Zh.I. Alferov, Phys. Rev. B **52**, 14058 (1995)

[44] C.D. Simserides, U. Hohenester, G. Goldone, E. Molinari, Phys. Rev. B **62**, 13657 (2000)

[45] T. Saiki, K. Matsuda, S. Nomura, M. Mihara, Y. Aoyagi, S. Nair, T. Takagahara, J. Electron Microsc. **53**, 193 (2004)

[46] Q. Wu, R.D. Grober, D. Gammon, D.S. Katzer, Phys. Rev. B **62**, 13022 (2000)

[47] D. Gammon, E.S. Snow, D.S. Katzer, Appl. Phys. Lett. **67**, 2391 (1995)

[48] S.V. Nair, T. Takagahara, Phys. Rev. B **55**, 5153 (1997)

[49] M.F. Crommie, C.P. Lutz, D.M. Eigler, Nature **363**, 524 (1993)

[50] M.F. Crommie, C.P. Lutz, D.M. Eigler, Science **262**, 218 (1993)

[51] M. Morgenstern, J. Klijn, C. Meyer, M. Getzlaff, R. Adelung, R.A. Romer, K. Rossnagel, L. Kipp, M. Skibowski, R. Wiesendanger, Phys. Rev. Lett. **89**, 136806 (2002)

[52] C. Meyer, J. Klijn, M. Morgenstern, R. Wiesendanger, Phys. Rev. Lett. **91**, 076803 (2003)

[53] K. Kanisawa, M.J. Butcher, Y. Tokura, H. Yamaguchi, Y. Hirayama, Phys. Rev. Lett. **87**, 196804 (2001)

[54] C. Chicanne, T. David, R. Quidant, J.C. Weeber, Y. Lacroute, E. Bourillot, A. Dereux, G. Colas des Francs, C. Girard, Phys. Rev. Lett. **88**, 097402 (2002)

[55] D. Heitmann, J.P. Kotthaus, Phys. Today **46**, 56 (1993)

[56] A. Lorke, J.P. Kotthaus, K. Ploog, Phys. Rev. Lett. **64**, 2559 (1990)

[57] W. Hansen, T.P. Smith III, K.Y. Lee, J.A. Brum, C.M. Knoedler, J.M. Hong, D.P. Kern, Phys. Rev. Lett. **62**, 2168 (1989)

[58] S. Nomura, Y. Aoyagi, Phys. Rev. Lett. **93**, 096803 (2004)

[59] V. Kukushkin, K. von Klitzing, K. Ploog, V.B. Timofeev, Phys. Rev. B **40**, 7788 (1989)

[60] V. Kukushkin, R.J. Haung, K. von Klitzing, K. Ploog, Phys. Rev. Lett. **72**, 736 (1994)

[61] S. Nomura, Y. Aoyagi, Surf. Sci. **529**, 171 (2003)

[62] S. Nomura, T. Nakanishi, Y. Aoyagi, Phys. Rev. B **63**, 165330 (2001)

[63] A. Kumar, S.E. Laux, F. Stern, Phys. Rev. B **42**, 5166 (1990)

[64] M.P. Stopa, Phys. Rev. B **54**, 13767 (1996)

[65] D. Gammon, E.S. Snow, B.V. Shanabrook, D.S. Katzer, D. Park, Science **273**, 87 (1996)

[66] M. Grundmann, J. Christen, N.N. Ledentsov, J. Bohrer, D. Bimberg, S.S. Ruvimov, P. Werner, U. Richter, U. Gosele, J. Heydenreich, V.M. Ustinov, A.Yu. Egorov, A.E. Zhukov, P.S. Kop'ev, Zh.I. Alferov, Phys. Rev. Lett. **74**, 4043 (1995)

[67] A. Buyanova, W.M. Chen, G. Pozina, J.P. Bergman, B. Monemar, H.P. Xin, C.W. Tu, Appl. Phys. Lett. **75**, 501 (1999)

[68] M. Kondow, K. Uomi, A. Niwa, T. Kitatani, S. Watahiki, Y. Yazawa, Jpn. J. Appl. Phys. **35**, 1273 (1996)

[69] D. Gollub, M. Fischer, M. Kamp, A. Forchel, Appl. Phys. Lett. **81**, 4330 (2002)

[70] M. Kawaguchi, T. Miyamoto, E. Gouardes, D. Schlenker, T. Kondo, F. Koyama, K. Iga, Jpn. J. Appl. Phys. **40**, L744 (2001)

[71] F. Masia, A. Polimeni, G.B.H. von Hogersthal, M. Bissiri, M. Capizzi, P.J. Klar, W. Stolz, Appl. Phys. Lett. **82**, 4474 (2003)

[72] S. Francoeur, S.A. Nikishin, C. Jin, Y. Qiu, H. Temkin, Appl. Phys. Lett. **75**, 1538 (1999)

[73] A. Moto, S. Tanaka, N. Ikoma, T. Tanabe, S. Takagishi, M. Takahashi, T. Katsuyama, Jpn. J. Appl. Phys. **38**, 1015 (1999)

[74] J. Sik, M. Schubert, G. Leibiger, V. Gottschalch, G. Wagner, J. Appl. Phys. **89**, 294 (2001)

[75] K. Matsuda, T. Saiki, S. Takahashi, A. Moto, M. Takagishi, Appl. Phys. Lett. **78**, 1508 (2001)

[76] M. Takahashi, A. Moto, S. Tanaka, T. Tanabe, S. Takagishi, K. Karatani, M. Nakayama, K. Matsuda, T. Saiki, J. Cryst. Growth **221**, 461 (2001)

[77] M. Mintairov, T.H. Kosel, J.L. Merz, P.A. Blagnov, A.S. Vlasov, V.M. Ustinov, R.E. Cook, Phys. Rev. Lett. **87**, 277401 (2001)

[78] L. Grenouillet, C. Bru-Chevallier, G. Guillot, P. Gilet, P. Duvaut, C. Vannuffel, A. Million, A. Chenevas-Paule, Appl. Phys. Lett. **76**, 2241 (2000)

[79] X.D. Luo, J.S. Huang, Z.Y. Xu, C.L. Yang, J. Liu, W.K. Ge, Y. Zhang, A. Mascarenhas, H.P. Xin, C.W. Tu, Appl. Phys. Lett. **82**, 1697 (2003)

[80] Y.J. Wang, X. Wei, Y. Zhang, A. Mascarenhas, H.P. Xin, Y.G. Hong, C.W. Tu, Appl. Phys. Lett. **82**, 4453 (2003)

2

Localized Photon Model Including Phonons' Degrees of Freedom

K. Kobayashi, Y. Tanaka, T. Kawazoe, and M. Ohtsu

2.1 Introduction

Optical near fields have been used in high-resolution microscopy/spectroscopy for a variety of samples [1], especially for a single molecule [2] and a single quantum dot [3], as well as for nanofabrication [4–6]. These applications are based on the fact that optical near-field probes, whose tips are sharpened to a few nanometers, can generate a light field localized around the apex of the same order. The spatial localization is, of course, independent of the wavelength of incident light, and the size of the localization is much smaller than the wavelength. It means that optical near-field probes are essential elements in these applications.

In fabricating nanophotonic devices [7–10] with such probes, for example, it is critical to control the size and position of the nanostructures, which requires efficient control and manipulation of the localization of light fields. If one could control and manipulate the localization of a light field at will, one would necessarily obtain more efficient and functional probes with higher precision, which will be applicable to predicting quantum phenomena. This is true not only in a probe system, but also in an optical near-field problem in general. In these respects it is very important to clarify the mechanism of spatial localization of optical near fields on a nanometer scale.

From a theoretical viewpoint, self-consistency between the light field and induced electronic polarization fields is crucial on the nanometer scale, and the importance of quantum coherence between photon and matter fields has been discussed [11–14]. On this basis, superradiance—as a cooperative phenomenon—of a quantum-dot chain system excited by an optical near field [15, 16], and excitation transfer to a dipole-forbidden level in a quantum-dot pair system [17, 18] have been investigated.

Recently, experimental results on superradiance using a collection of quantum dots have been reported [19]. Moreover, experiments on photodissociation of diethylzinc (DeZn) and zinc-bis (acetylacetonate) or $Zn(acac)_2$ molecules and deposition of Zn atoms using an optical near field have been conducted for nanostructure fabrication, as discussed in the next section. The experimental results show that the molecules illuminated by the optical near field are dissociated even if the energy of incident light is lower than the dissociation energy, which is impossible when a far field with the same energy and intensity is used. A simple analysis indicates that data cannot be explained by conventional theories based on the Franck–Condon principle or the adiabatic approximation for nuclear motion in a molecule, and suggests that phonons in an optically excited probe system might assist the molecular dissociation process in a nonadiabatic way [20–22].

In this situation, it is necessary to study the photon–phonon interaction as well as the photon–electronic excitation interaction in a nanometer space, and to clarify the phonon's role in the nanometric optical near-field probe system, or more generally in light-matter interacting system on a nanometer scale. Then a quantum theoretical approach is appropriate to describe an interacting system of photon and matter (electronic excitation and phonon) fields. It allows us not only to understand an elementary process of photochemical reactions with optical near fields, but also to explore the role that phonons play in nanostructures interacting with localized photon fields.

2.2 Quantum Theoretical Approach to Optical Near Fields

A "photon", as is well known, corresponds to a discrete excitation of electromagnetic modes in a virtual cavity, whose concept has been established as a result of quantization of a free electromagnetic field (see, for example, [23]). Different from an electron, a photon is massless, and it is hard to construct a wave function in the coordinate representation that gives a photon picture as a spatially localized point particle as an electron [24]. However if there is a detector, such as an atom, to absorb a photon in an area whose linear dimension is much smaller than the wavelength of light, it would be possible to detect a photon with the same precision as the detector size [25, 26]. In optical near-field problems, we are required to consider the interactions between light and nanomaterials and detection of light by other nanomaterials on a nanometer scale. Then it is more serious for quantization of the field regarding how to define a virtual cavity, or which normal modes are to be used, since there exists a system composed of separated materials with arbitrary size and shape on the nanometer region.

In this section, we describe a theoretical approach to address the issue. Then we discuss photodissociation of molecules as an example of applications using optical near fields, which is an essential part of nanofabrication to construct nanophotonic devices.

2.2.1 Localized Photon Model

Effective Interaction and Localized Photons

Let us consider a nanomaterial system surrounded by an incident light and a macroscopic material system, which are electromagnetically interacting with one another

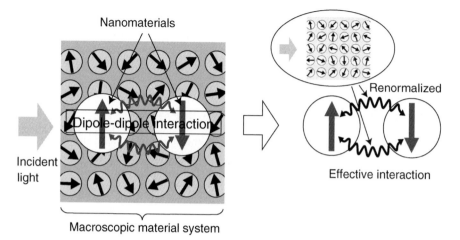

Fig. 2.1. Schematic drawing of near-field optical interactions. The effects from the irrelevant system, in which we are not interested, are renormalized as the effective interaction between nanomaterials

in a complicated way, as shown in Fig. 2.1. Using the projection operator method, we can derive an effective interaction between the relevant nanomaterials in which we are interested, after renormalizing the other effects [9, 13, 27, 28]. It corresponds to an approach describing "photons localized around nanomaterials" as if each nanomaterial would work as a detector and light source in a self-consistent way. The effective interaction related to optical near fields is hereafter called a near-field optical interaction. As discussed in [9, 13, 27, 28] in detail, the near-field optical interaction potential between nanomaterials separated by R is given as

$$V_{\text{eff}} = \frac{\exp(-aR)}{R}, \tag{2.1}$$

where a^{-1} is the interaction range that represents the characteristic size of nanomaterials and does not depend on the wavelength of light. It indicates that photons are localized around the nanomaterials as a result of the interaction with matter fields, from which a photon, in turn, can acquire a finite mass. Therefore we might consider if the near-field optical interaction would be produced via the localized photon hopping [15, 29, 30] between nanomaterials.

In an example, let us look at an optically excited probe system whose apex is sharpened on a nanometer scale, where the radius of curvature of the probe tip, r_0, is regarded as a characteristic size of the light-matter interacting system. The probe system is coarse-grained in terms of r_0, and then photons are localized at each coarse-grained point with interaction range r_0, which causes the localized photons to hop the nearest neighbor points.

In experiments, the above explanation can be applied to usual near-field imaging and spectroscopy. In addition, the near-field optical interactions between semiconductor quantum dots [17, 31, 32], between semiconductor nanorods [33], as well as light-harvesting antenna complex of photosynthetic purple bacteria [34, 35], have been observed by using the optically forbidden excitation energy transfer. Moreover, the near-field optical interactions are used in nanofabrication, which will be discussed in the following section.

2.2.2 Photodissociation of Molecules and the EPP Model

We outline the nanofabrication technique using the optical near field, and discuss the unique feature found in the results of photodissociation experiments, on the basis of a simple model—the exciton–phonon polariton (EPP) model.

Experimental

As illustrated in Fig. 2.2, optical near-field chemical vapor deposition (NFO-CVD) is used to fabricate a nanometer-scale structure while controlling position and size [4, 5]. Incident laser light is introduced into an optical near-field probe, i.e., a glass fiber that is chemically etched to have a nanometric sized apex without the metal coating usually employed. The propagating far field is generated by light leaking through the circumference of the fiber, while the optical near field is mainly generated at the apex. This allows us to investigate the deposition by an optical near field and far field simultaneously. The separation between the fiber probe and the sapphire (0001) substrate is kept within a few nanometers by shear-force feedback control. By appropriately selecting reactant molecules to be dissociated, NFO-CVD is applicable to various materials such as metals, semiconductors and insulators. In the following, however, we concentrate on diethylzinc (DEZn) and zinc-bis (acetylacetonate) $(Zn(acac)_2)$ as reactant molecules, at 70–100 mTorr at room temperature.

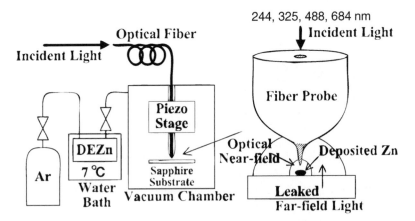

Fig. 2.2. Experimental setup for chemical vapor deposition using an optical near field. The DEZn bottle and CVD chamber were kept at 7 and 25 degrees C, respectively, to prevent the condensation of DEZn on the sapphire substrate. During deposition, the partial pressure of DEZn was 100 mTorr and the total pressure in the chamber was 8 Torr

In order to investigate the mechanism of the photochemical process, deposition rates depending on photon energy and intensity have been measured with several laser sources. For DEZn molecules:

1. The second harmonic of an Ar^+ laser ($\hbar\omega = 5.08$ eV, corresponding wavelength $\lambda = 244$ nm), whose energy is close to the electronic excitation energy (5 eV) of a DEZn molecule.
2. An He-Cd laser ($\hbar\omega = 3.81$ eV, corresponding wavelength $\lambda = 325$ nm), whose energy is close to $E_{abs} \sim 4.13$ eV [36, 37] corresponding to the energy of the absorption band edge.
3. An Ar^+ laser ($\hbar\omega = 2.54$ eV, corresponding wavelength $\lambda = 488$ nm), whose energy is larger than the dissociation energy of the molecule (2.26 eV), but much smaller than the electronic excitation energy and E_{abs}.
4. A diode laser ($\hbar\omega = 1.81$ eV, corresponding wavelength $\lambda = 684$ nm), whose energy is smaller than both the dissociation and electronic excitation energies, as well as E_{abs}.

And for Zn(acac)$_2$:

1. An Ar^+ laser ($\hbar\omega = 2.71$ eV, corresponding wavelength $\lambda = 457$ nm), whose energy is much smaller than the electronic excitation energy and $E_{abs} \sim 5.17$ eV.

Shear-force topographical images are shown in Fig. 2.3 after NFO-CVD at the photon energies listed above. In conventional CVD using a propagating light, photon

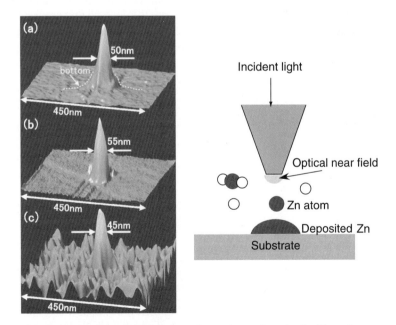

Fig. 2.3. Schematic drawing of NFO-CVD and experimental results. Incident photon energies are **a** 3.81 eV, **b** 2.54 eV and **c** 1.81 eV

energy must be higher than E_{abs} because dissociative molecules should be excited from the ground state to an excited electronic state, according to the adiabatic approximation [38, 39]. In contrast, even if photon energy less than E_{abs} is employed in NFO-CVD, the deposition of Zn dots are observed on the substrate just below the apex of the probe used. Much more interestingly, photons with less energy than the dissociation energy can resolve both DEZn and Zn(acac)$_2$ molecules into composite atoms and deposit them as nanometric dots [20, 40].

One possibility inferred from the results is a multiple photon absorption process, which is negligibly small because the optical power density used in the experiment was less than 10 kW/cm^2; that is too low for the process. The other possibility is a multiple step transition via an excited molecular vibrational level that is forbidden by the Frank–Condon principle, but allowed in a nonadiabatic process. In order to clarify the unique feature of NFO-CVD, in the following we give a simple model to discuss the process.

EPP Model

We propose a quasiparticle (exciton–phonon polariton) model as a simple model of an optically excited probe system, in order to investigate the physical mechanisms of the chemical vapor deposition using an optical near field (NFO-CVD) [21]. We assume that exciton–phonon polaritons, the quanta of which are transferred from the optical near-field probe tip to both gas and adsorbed molecules, are created at the apex of the optical near-field probe. Here it should be noted that the quasi-particle transfer is valid only if the molecules are very close to the probe tip because the optical near field is highly localized near the probe tip, which is discussed in Sect. 2.4.4. The optical near field generated on the nanometric probe tip, which is a highly mixed state with material excitation rather than the propagating light field [13, 28], is described in terms of the following model Hamiltonian:

$$
\begin{aligned}
H = & \sum_p \hbar \left\{ \omega_p a_p^\dagger a_p + \omega_p^{ex} b_p^\dagger b_p + \frac{i\Omega_c}{2} \left(a_p^\dagger b_p - b_p^\dagger a_p \right) \right\} \\
& + \sum_p \hbar \Omega_p c_p^\dagger c_p + \sum_{p,q} \left[i\hbar M(p-q) b_p^\dagger b_q \{ c_{p-q} + c_{q-p}^\dagger \} + h.c. \right] \\
= & \sum_p \hbar \omega_p^{pol} B_p^\dagger B_p + \sum_p \hbar \Omega_p c_p^\dagger c_p \\
& + \sum_{p,q} \left[i\hbar M'(p-q) B_p^\dagger B_q \{ c_{p-q} + c_{q-p}^\dagger \} + h.c. \right],
\end{aligned}
\tag{2.2}
$$

where the creation (annihilation) operators for a photon, an exciton (a quasiparticle for an electronic polarization field), a renormalized phonon (whose physical meanings are discussed in Sect. 2.4.3), and an exciton polariton are denoted by $a_p^\dagger$ (a_p), $b_p^\dagger$ (b_p), $c_p^\dagger$ (c_p), and $B_p^\dagger$ (B_p), respectively, and their frequencies are ω_p, ω_p^{ex}, Ω_p, and ω_p^{pol}, respectively. The subscripts p and q indicate the momenta of the relevant particles in the momentum representation such as a photon, an exciton, a renormalized phonon, an exciton polariton, or an exciton–phonon polariton. Each coupling between a photon and an exciton, a phonon and an exciton, and an exciton polariton and a phonon is designated as Ω_c, $M(p - q)$, and $M'(p - q)$, respectively. The first line of this description expresses the Hamiltonian for a photon–exciton interacting system and is transformed into the exciton–polariton representation as shown in the third line [41], while the second line represents the Hamiltonian for a phonon–exciton interacting system. Note that electronic excitations near the probe tip, driven by photons incident into the fiber probe, cause mode–mode couplings or anharmonic couplings of phonons, and that they are taken into account as a renormalized phonon; therefore, multiple phonons as coherently squeezed phonons in the original representation can interact with an exciton or an exciton polariton simultaneously. In the model, quasi-particles (exciton–phonon polaritons) in bulk material (glass fiber) are approximately used, and thus their states are specified by the momentum. Strictly speaking, momentum is not a good quantum number to specify the quasi-particle states at the apex of the probe, from the symmetry consideration, and they should be a superposition of such momentum-specified states with different weights. Instead of this kind of treatment, we simply assume that the quasi particles specified by the momentum are transferred to a vapor or adsorbed molecule that is located near the probe tip, using highly spatial localization of the optical near field to be discussed in Sect. 2.4.4 in detail.

Now we assume that exciton polaritons near the probe tip are expressed in the mean field approximation as

$$\langle B_{k_0}^\dagger \rangle = \langle B_{k_0} \rangle = \sqrt{\frac{I_0(\omega_0)V}{\hbar\omega_0 d}}. \tag{2.3}$$

Here $I_0(\omega_0)$ is the photon intensity inside the probe tip with frequency ω_0 and momentum $\hbar k_0$, and V represents the volume to be considered while the probe tip size is denoted by d. Using the unitary transformation as

$$\begin{pmatrix} B_p \\ c_{p-k_0} \end{pmatrix} = \begin{pmatrix} iv'_p & u'_p \\ u'_p & iv'_p \end{pmatrix} \begin{pmatrix} \xi_{(-)p} \\ \xi_{(+)p} \end{pmatrix}, \tag{2.4}$$

we can diagonalize the Hamiltonian in the exciton–phonon polariton representation [42] as

$$
\begin{aligned}
H &= \sum_p \hbar\omega_p^{\text{pol}} B_p^\dagger B_p + \sum_p \hbar\Omega_p c_p^\dagger c_p \\
&\quad + \sum_p \left\{ i\hbar\sqrt{\frac{I_0(\omega_0)V}{\hbar\omega_0 d}} M'(p - k_0)\left(B_p^\dagger c_{p-k_0} - B_p c_{p-k_0}^\dagger\right) \right\} \\
&= \sum_p \sum_{j=\pm} \hbar\omega(p) \xi_{jp}^\dagger \xi_{jp},
\end{aligned} \tag{2.5}
$$

where the creation (annihilation) operator for an exciton–phonon polariton and the frequency are denoted by $\xi_{jp}^\dagger (\xi_{jp})$ and $\omega(p)$, respectively. The suffix $(-)$ or $(+)$

indicates the lower or upper branch of the exciton–phonon polariton. The transformation coefficients u'_p and v'_p are given by

$$u'^2_p = \frac{1}{2}\left(1 + \frac{\Delta}{\sqrt{\Delta^2 + (2Q)^2}}\right), \qquad v'^2_p = \frac{1}{2}\left(1 - \frac{\Delta}{\sqrt{\Delta^2 + (2Q)^2}}\right), \qquad (2.6)$$

where the detuning between an exciton polariton and a phonon is denoted by $\Delta = \omega_p^{\mathrm{pol}} - \Omega_{p-k_0}$, and the effective coupling constant is expressed as $Q = \sqrt{I_0(\omega_0)V/(\hbar\omega_0 d)}M'(p - k_0)$. Therefore, in this model, a molecule located near the probe tip does absorb not simple photons but exciton–phonon polaritons whose energies are transferred to the molecule, which excite molecular vibrations as well as electronic transitions to elucidate the experimental results. In the following sections, we discuss how phonons work in the optically excited probe system in detail.

2.3 Localized Phonons

In this section, lattice vibrations in a pseudo one-dimensional system are briefly described and then quantized. We examine the effects of impurities or defects in such a system to show that the localized vibration modes exist as eigenmodes, and those energies are higher than those of delocalized ones.

2.3.1 Lattice Vibration in a Pseudo One-Dimensional System

Owing to the progress in nanofabrication, the apexes of optical near-field probes are sharpened on the order of a few nanometers. In this region, the guiding modes of light field are cut off and visible light cannot propagate in a conventional way. Therefore it is necessary to clarify the interactions among light, induced electronic, and vibrational fields on the nanometer space—such as the optically excited probe tip—and the mechanism of localization (delocalization) of light field as a result of self-consistency of those interacting fields. As the first step, we examine the lattice vibrations themselves in this section.

Let us assume a pseudo one-dimensional system for the probe tip, as illustrated in Fig. 2.4. The system consists of a finite number (N) of atoms or molecules, which will be representatively called molecules. Each molecule is located at a discrete site and is connected with the nearest-neighbor molecules by springs. The size of each molecule and the spacing between the molecules depend on how the system is coarse-grained. In any case, the total site number N is finite, and the wave number is not a good quantum number because the system breaks the translational invariance [43]. That is why we begin with the Hamiltonian of the system to analyze vibrational (phonon) modes, instead of the conventional method using the dynamical matrix [44]. Denoting a displacement from an equilibrium point of a molecule at site i as x_i and its conjugate momentum by p_i, we write the model Hamiltonian as

$$H = \sum_{i=1}^{N} \frac{p_i^2}{2m_i} + \sum_{i=1}^{N-1} \frac{1}{2}k\,(x_{i+1} - x_i)^2 + \frac{1}{2}kx_1^2 + \frac{1}{2}kx_N^2, \qquad (2.7)$$

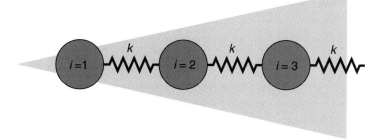

Fig. 2.4. A pseudo one-dimensional system for a NFO probe tip

where m_i is the mass of a molecule at site i, and k represents the spring constant. Both edges ($i = 1$ and $i = N$) are assumed to be fixed, and longitudinal motions in one-dimension are considered in the following.

The equations of motion are determined by the Hamilton equation as

$$\frac{\mathrm{d}}{\mathrm{d}t}x_i = \frac{\partial H}{\partial p_i}, \qquad \frac{\mathrm{d}}{\mathrm{d}t}p_i = -\frac{\partial H}{\partial x_i}. \qquad (2.8)$$

If one uses a matrix form defined by

$$\mathbf{M} = \begin{pmatrix} m_1 & & & \\ & m_2 & & \\ & & \ddots & \\ & & & m_N \end{pmatrix}, \qquad \mathbf{\Gamma} = \begin{pmatrix} 2 & -1 & & \\ -1 & 2 & \ddots & \\ & \ddots & \ddots & -1 \\ & & -1 & 2 \end{pmatrix}, \qquad (2.9)$$

one can obtain the following compact equations of motion:

$$\mathbf{M}\frac{\mathrm{d}^2}{\mathrm{d}t^2}x = -k\mathbf{\Gamma}x, \qquad (2.10)$$

with transpose of the column vector x as

$$x^{\mathrm{T}} \equiv (x_1, x_2, \ldots, x_N). \qquad (2.11)$$

Multiplying the both sides of (2.10) by $\sqrt{\mathbf{M}}^{-1}$ with $(\sqrt{\mathbf{M}})_{ij} = \delta_{ij}\sqrt{m_i}$, we have

$$\frac{\mathrm{d}^2}{\mathrm{d}t^2}x' = -k\mathbf{A}x', \qquad (2.12)$$

where the notations $x' = \sqrt{\mathbf{M}}x$ and $\mathbf{A} = \sqrt{\mathbf{M}}^{-1}\mathbf{\Gamma}\sqrt{\mathbf{M}}^{-1}$ are used. Since it is symmetric, the matrix $\mathbf{A}$ can be diagonalized by an orthonormal matrix $\mathbf{P}$ as follows:

$$\mathbf{\Lambda} = \mathbf{P}^{-1}\mathbf{A}\mathbf{P}, \quad \text{or} \quad (\mathbf{\Lambda})_{pq} = \delta_{pq}\frac{\Omega_p^2}{k}. \qquad (2.13)$$

Substitution of (2.13) into (2.12) leads us to equations of motion for a set of harmonic oscillators as

$$\frac{d^2}{dt^2}\mathbf{y} = -k\Lambda\mathbf{y}, \quad \text{or} \quad \frac{d^2}{dt^2}y_p = -\Omega_p^2 y_p, \tag{2.14}$$

where $\mathbf{y}$ is set as $\mathbf{y} = \mathbf{P}^{-1}\mathbf{x}'$. There are N normal coordinates to describe the harmonic oscillators, each of which is specified by the mode number p. The original spatial coordinates $\mathbf{x}$ are transformed to the normal coordinates $\mathbf{y}$ as

$$\mathbf{x} = \sqrt{\mathbf{M}}^{-1}\mathbf{P}\mathbf{y}, \quad \text{or} \quad x_i = \frac{1}{\sqrt{m_i}}\sum_{p=1}^{N} P_{ip} y_p. \tag{2.15}$$

2.3.2 Quantization of Vibration

In order to quantize the vibration field described by (2.14), we first rewrite the Hamiltonian (2.7) in terms of normal coordinates $\hat{y}_p$ and conjugate momenta $\hat{\pi}_p$ as

$$H(\hat{y}, \hat{\pi}) = \sum_{p=1}^{N} \frac{1}{2}\hat{\pi}_p^2 + \sum_{p=1}^{N} \frac{1}{2}\Omega_p^2 \hat{y}_p^2. \tag{2.16}$$

Then the commutation relation between $\hat{y}_p$ and $\hat{\pi}_q$ as

$$[\hat{y}_p, \hat{\pi}_q] = i\hbar\delta_{pq}, \tag{2.17}$$

is imposed for quantization. When we define operators $\hat{b}_p$ and $\hat{b}_p^\dagger$ as

$$\hat{b}_p = \frac{1}{\sqrt{2\hbar\Omega_p}}(\hat{\pi}_p - i\Omega_p\hat{y}_p), \tag{2.18a}$$

$$\hat{b}_p^\dagger = \frac{1}{\sqrt{2\hbar\Omega_p}}(\hat{\pi}_p + i\Omega_p\hat{y}_p), \tag{2.18b}$$

they satisfy the boson commutation relation

$$[\hat{b}_p, \hat{b}_q^\dagger] = \delta_{pq}. \tag{2.19}$$

The Hamiltonian describing the lattice vibration of the system, (2.16), can then be rewritten as

$$\hat{H}_{\text{phonon}} = \sum_{p=1}^{N} \hbar\Omega_p\left(\hat{b}_p^\dagger\hat{b}_p + \frac{1}{2}\right), \tag{2.20}$$

and it follows that $\hat{b}_p$ ($\hat{b}_p^\dagger$) is the annihilation (creation) operator of a phonon with energy of $\hbar\Omega_p$ specified by the mode number p.

2.3.3 Vibration Modes: Localized vs. Delocalized

In this section, we examine the effects of impurities or defects in the system. When all the molecules are identical, i.e., $m_i = m$, the Hamiltonian (2.7), or the matrix $\mathbf{A}$ can be diagonalized in terms of the orthonormal matrix $\mathbf{P}$ whose elements are given by

$$\mathbf{P}_{ip} = \sqrt{\frac{2}{N+1}} \sin\left(\frac{ip}{N+1}\pi\right) \quad (1 \le i, p \le N), \tag{2.21}$$

and the eigenfrequencies squared are obtained as follows:

$$\Omega_p^2 = 4\frac{k}{m} \sin^2\left(\frac{p}{2(N+1)}\pi\right). \tag{2.22}$$

In this case, all the vibration modes are delocalized, i.e., they are spread over the whole system. On the other hand, if there are some doped impurities or defects with different mass, the vibration modes depend highly on the geometrical configuration and mass ratio of the impurities to the others. In particular, localized vibration modes manifest themselves when the mass of the impurities is lighter than that of the others, where vibrations with higher frequencies are localized around the impurity sites [45–48].

Figures 2.5(a) and (b) illustrate that the localized vibration modes exist as eigen-modes in the one-dimensional system due to the doped molecules with different mass in the chain, and eigenenergies of localized modes are higher than those of delo-calized ones. In Fig. 2.5(a), phonon energies are plotted as a function of the mode number when the total number of sites is 30. The squares represent the eigenenergies of phonons in the case of no impurities, and the circles show those in the case of six impurities, where the doped molecules are located at sites 5, 9, 18, 25, 26, 27. It follows from the figure that phonon energies of the localized modes are higher than those of the delocalized modes. The mass ratio of the doped molecules to the others is $1/2$, and the parameter $\hbar\sqrt{k/m} = 22.4\,\text{meV}$ is used. Figure 2.5(b) shows the vi-bration amplitude as a function of the site number. The solid curve with squares and the dashed curve with circles represent two localized modes with the highest and the next highest energies of phonons, respectively, while the dotted curve with triangles illustrates the delocalized mode with the lowest energy. In the localized modes, the vibration amplitudes are localized around the impurity sites.

In the next section, we discuss the interactions between photons and inhomoge-neous phonon fields on the nanometer scale, since we have found inhomogeneous phonon fields in the one-dimensional system with impurities.

2.4 Extended Model

In this section, we propose a simple model for a pseudo one-dimensional optical near-field probe system to discuss the mechanism of photon localization in space as

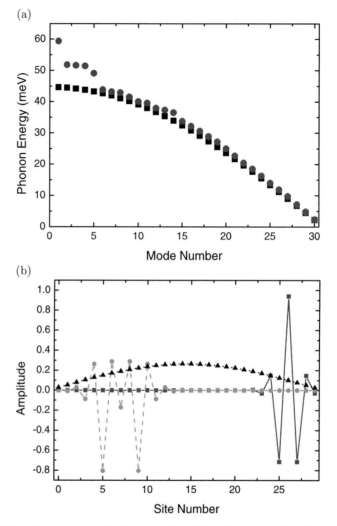

Fig. 2.5. a Eigenfrequencies of all phonon modes with/without impurities (depicted with the circles/squares), in the case of $N = 30$, and **b** the first and second localized phonon modes and the lowest delocalized phonon mode (represented by the solid, dashed and dotted curves). Impurities are doped at sites 5, 9, 18, 25, 26 and 27. The mass ratio of the host molecules to the impurities is set as 1 to 0.5, and $\hbar\sqrt{k/m} = 22.4\,\text{meV}$ is used for both **a** and **b**

well as the role of phonons. In order to focus on the photon–phonon interaction, the interacting part between photon and electronic excitation is first expressed in terms of a polariton, and is called a photon in the model. Then the model Hamiltonian, which describes the photon and phonon interacting system, is presented. Using the Davydov transformation [43, 49, 50], we rewrite the Hamiltonian in terms of qua-siparticles. On the basis of the Hamiltonian, we present numerical results on spatial distribution of photons and discuss the mechanism of photon localization due to phonons.

2.4.1 Optically Excited Probe System

We consider an optical near-field probe, schematically shown in Fig. 2.4, as a system where light interacts with both phonons and electrons in the probe on a nanometer scale. Here the interaction of a photon and an electronic excitation is assumed to be expressed in terms of a polariton basis [28] as discussed in Sect. 2.2.2, and is hereafter called a photon so that special attention is paid to the photon–phonon interaction. The system is simply modeled as a one-dimensional atomic or molecular chain coupled with photon and phonon fields. The chain consists of a finite N molecules (representatively called) each of which is located at a discrete point (called a molecular site) whose separation represents a characteristic scale of the near-field system. Photons are expressed in the site representation and can hop to the nearest neighbor sites [15] due to the short-range interaction nature of the optical near fields (see (2.1)).

The Hamiltonian for the above model is given by

$$
\hat{H} = \sum_{i=1}^{N} \hbar\omega \hat{a}_i^\dagger \hat{a}_i
$$
$$
+ \left\{ \sum_{i=1}^{N} \frac{\hat{p}_i^2}{2m_i} + \sum_{i=1}^{N-1} \frac{k}{2}(\hat{x}_{i+1} - \hat{x}_i)^2 + \sum_{i=1,N} \frac{k}{2}\hat{x}_i^2 \right\}
$$
$$
+ \sum_{i=1}^{N} \hbar\chi \hat{a}_i^\dagger \hat{a}_i \hat{x}_i + \sum_{i=1}^{N-1} \hbar J \left(\hat{a}_i^\dagger \hat{a}_{i+1} + \hat{a}_{i+1}^\dagger \hat{a}_i \right), \tag{2.23}
$$

where $\hat{a}_i^\dagger$ and $\hat{a}_i$ correspondingly denote the creation and annihilation operators of a photon with energy of $\hbar\omega$ at site i in the chain, and $\hat{x}_i$ and $\hat{p}_i$ represent the displacement and conjugate momentum operators of the vibration, respectively. The mass of a molecule at site i is designated by m_i, and each molecule is assumed to be connected by springs with spring constant k. The third and the fourth terms in (2.23) stand for the photon–vibration interaction with coupling constant χ and the photon hopping with hopping constant J, respectively.

After the vibration field is quantized in terms of phonon operators of mode p and frequency Ω_p, $\hat{b}_p^\dagger$ and $\hat{b}_p$, the Hamiltonian (2.23) can be rewritten as

$$
\hat{H} = \sum_{i=1}^{N} \hbar\omega \hat{a}_i^\dagger \hat{a}_i + \sum_{p=1}^{N} \hbar\Omega_p \hat{b}_p^\dagger \hat{b}_p
$$
$$
+ \sum_{i=1}^{N} \sum_{p=1}^{N} \hbar\chi_{i,p} \hat{a}_i^\dagger \hat{a}_i \left(\hat{b}_p^\dagger + \hat{b}_p \right)
$$
$$
+ \sum_{i=1}^{N-1} \hbar J \left(\hat{a}_i^\dagger \hat{a}_{i+1} + \hat{a}_{i+1}^\dagger \hat{a}_i \right), \tag{2.24}
$$

with the coupling constant $\chi_{i,p}$ between a photon at site i and a phonon of mode p. This site-dependent coupling constant $\chi_{i,p}$ is related to the original coupling constant χ as

$$\chi_{i,p} = \chi \mathbf{P}_{ip}\sqrt{\frac{\hbar}{2m_i \Omega_p}}, \tag{2.25}$$

and the creation and annihilation operators of a photon and a phonon satisfy the boson commutation relation as follows:

$$
\begin{aligned}
&[\hat{a}_i, \hat{a}_j^\dagger] = \delta_{ij}, \qquad [\hat{b}_p, \hat{b}_q^\dagger] = \delta_{pq}, \\
&[\hat{a}_i, \hat{a}_j] = [\hat{a}_i^\dagger, \hat{a}_j^\dagger] = 0 = [\hat{b}_p, \hat{b}_q] = [\hat{b}_p^\dagger, \hat{b}_q^\dagger], \\
&[\hat{a}_i, \hat{b}_p] = [\hat{a}_i, \hat{b}_p^\dagger] = [\hat{a}_i^\dagger, \hat{b}_p] = [\hat{a}_i^\dagger, \hat{b}_p^\dagger] = 0.
\end{aligned} \tag{2.26}
$$

The Hamiltonian (2.24), which describes the model system, is not easily handled because of the third order of the operators in the interaction term. To avoid the difficulty, this direct photon–phonon interaction term in (2.24) will be eliminated by the Davydov transformation in the following section.

2.4.2 Davydov Transformation

Before going into the explicit expression, we discuss a unitary transformation $\hat{U}$ generated by an anti-Hermitian operator $\hat{S}$ defined as

$$
\begin{aligned}
\hat{U} &\equiv e^{\hat{S}}, \quad \text{with } \hat{S}^\dagger = -\hat{S}, \tag{2.27a} \\
\hat{U}^\dagger &= \hat{U}^{-1}. \tag{2.27b}
\end{aligned}
$$

Suppose a Hamiltonian $\hat{H}$ that consists of a diagonalized part $\hat{H}_0$ and a nondiagonal interaction part $\hat{V}$ as

$$\hat{H} = \hat{H}_0 + \hat{V}. \tag{2.28}$$

Transforming the Hamiltonian in (2.28) as

$$\tilde{H} \equiv \hat{U}\hat{H}\hat{U}^\dagger = \hat{U}\hat{H}\hat{U}^{-1}, \tag{2.29}$$

we have

$$
\begin{aligned}
\tilde{H} &= \hat{H} + [\hat{S}, \hat{H}] + \frac{1}{2}[\hat{S}, [\hat{S}, \hat{H}]] + \cdots \\
&= \hat{H}_0 + \hat{V} + [\hat{S}, \hat{H}_0] + [\hat{S}, \hat{V}] + \frac{1}{2}[\hat{S}, [\hat{S}, \hat{H}_0]] + \cdots. \tag{2.30}
\end{aligned}
$$

If the interaction $\hat{V}$ can be perturbative, and if the operator $\hat{S}$ is chosen so that the second and the third terms in (2.30) are canceled out as

$$\hat{V} = -[\hat{S}, \hat{H}_0], \tag{2.31}$$

the Hamiltonian (2.30) is rewritten as

$$\tilde{H} = \hat{H}_0 - \frac{1}{2}[\hat{S}, [\hat{S}, \hat{H}_0]] + \cdots, \tag{2.32}$$

and can be diagonalized within the first order of $\hat{V}$.

Now we apply the above discussion to the model Hamiltonian (2.24),

$$\hat{H}_0 = \sum_{i=1}^{N} \hbar\omega\hat{a}_i^\dagger\hat{a}_i + \sum_{p=1}^{N} \hbar\Omega_p\hat{b}_p^\dagger\hat{b}_p, \tag{2.33a}$$

$$\hat{V} = \sum_{i=1}^{N}\sum_{p=1}^{N} \hbar\chi_{i,p}\hat{a}_i^\dagger\hat{a}_i(\hat{b}_p^\dagger + \hat{b}_p), \tag{2.33b}$$

tentatively neglecting the hopping term. Assuming the anti-Hermitian operator $\hat{S}$ as

$$\hat{S} = \sum_i\sum_p f_{ip}\hat{a}_i^\dagger\hat{a}_i(\hat{b}_p^\dagger - \hat{b}_p), \tag{2.34}$$

we can determine f_{ip} from (2.31) as follows:

$$f_{ip} = \frac{\chi_{ip}}{\Omega_p}. \tag{2.35}$$

This operator form of $\hat{S}$ leads us not to the perturbative but to the exact transformation of the photon and phonon operators as

$$\hat{\alpha}_i^\dagger \equiv \hat{U}^\dagger\hat{a}_i^\dagger\hat{U} = \hat{a}_i^\dagger\exp\left\{-\sum_{p=1}^{N}\frac{\chi_{ip}}{\Omega_p}(\hat{b}_p^\dagger - \hat{b}_p)\right\}, \tag{2.36a}$$

$$\hat{\alpha}_i \equiv \hat{U}^\dagger\hat{a}_i\hat{U} = \hat{a}_i\exp\left\{\sum_{p=1}^{N}\frac{\chi_{ip}}{\Omega_p}(\hat{b}_p^\dagger - \hat{b}_p)\right\}, \tag{2.36b}$$

$$\hat{\beta}_p^\dagger \equiv \hat{U}^\dagger\hat{b}_p^\dagger\hat{U} = \hat{b}_p^\dagger + \sum_{i=1}^{N}\frac{\chi_{ip}}{\Omega_p}\hat{a}_i^\dagger\hat{a}_i, \tag{2.36c}$$

$$\hat{\beta}_p \equiv \hat{U}^\dagger\hat{b}_p\hat{U} = \hat{b}_p + \sum_{i=1}^{N}\frac{\chi_{ip}}{\Omega_p}\hat{a}_i^\dagger\hat{a}_i. \tag{2.36d}$$

These transformed operators can be regarded as the creation and annihilation operators of quasiparticles—dressed photons and phonons—that satisfy the same boson commutation relations as those of photons and phonons before the transformation:

$$[\hat{\alpha}_i, \alpha_j^\dagger] = \hat{U}^\dagger[\hat{a}_i, a_j^\dagger]\hat{U} = \delta_{ij}, \tag{2.37a}$$

$$[\hat{\beta}_p, \beta_q^\dagger] = \hat{U}^\dagger[\hat{b}_p, b_q^\dagger]\hat{U} = \delta_{pq}. \tag{2.37b}$$

Using the quasiparticle operators, we can rewrite the Hamiltonian (2.24) as

$$
\hat{H} = \sum_{i=1}^{N} \hbar \omega \hat{\alpha}_i^\dagger \hat{\alpha}_i + \sum_{p=1}^{N} \hbar \Omega_p \hat{\beta}_p^\dagger \hat{\beta}_p - \sum_{i=1}^{N} \sum_{j=1}^{N} \sum_{p=1}^{N} \hbar \frac{\chi_{ip} \chi_{jp}}{\Omega_p} \hat{\alpha}_i^\dagger \hat{\alpha}_i \hat{\alpha}_j^\dagger \hat{\alpha}_j
$$

$$
+ \sum_{i=1}^{N-1} \hbar \left(\hat{J}_i \hat{\alpha}_i^\dagger \hat{\alpha}_{i+1} + \hat{J}_i^\dagger \hat{\alpha}_{i+1}^\dagger \hat{\alpha}_i \right), \tag{2.38}
$$

with

$$
\hat{J}_i = J \exp \left\{ \sum_{p=1}^{N} \frac{(\chi_{i,p} - \chi_{i+1,p})}{\Omega_p} (\hat{\beta}_p^\dagger - \hat{\beta}_p) \right\}, \tag{2.39}
$$

where it is noted that the direct photon–phonon coupling term has been eliminated while the quadratic form $\hat{N}_i \hat{N}_j$ with the number operator of $\hat{N}_i = \hat{\alpha}_i^\dagger \alpha_i$ has emerged as well as the site-dependent hopping operator (2.39). The number states of quasiparticles are thus eigenstates of each terms of the Hamiltonian (2.38), except the last term that represents the higher order effect of photon–phonon coupling through the dressed photon hopping. Therefore it is more appropriate to discuss the phonon's effect on the photon's behavior as localization.

2.4.3 Quasiparticle and Coherent State

In the previous section, we transformed the original Hamiltonian by the Davydov transformation. In order to grasp the physical meanings of the quasiparticles introduced above, the creation operator $\hat{\alpha}_i^\dagger$ is applied to the vacuum state $|0\rangle$. Then it follows from (2.36a)

$$
\hat{\alpha}_i^\dagger |0\rangle = \hat{a}_i^\dagger \exp \left\{ - \sum_{p=1}^{N} \frac{\chi_{ip}}{\Omega_p} (\hat{b}_p^\dagger - \hat{b}_p) \right\} |0\rangle,
$$

$$
= \hat{a}_i^\dagger \exp \left\{ - \sum_{p=1}^{N} \frac{1}{2} \left(\frac{\chi_{ip}}{\Omega_p} \right)^2 \right\} \exp \left\{ - \sum_{p=1}^{N} \frac{\chi_{ip}}{\Omega_p} \hat{b}_p^\dagger \right\} |0\rangle, \tag{2.40}
$$

where a photon at site i is associated with phonons in a coherent state, i.e., a photon is dressed by an infinite number of phonons. This corresponds to the fact that an optical near field is generated from a result of interactions between the photon and matter fields.

When $\beta_p^\dagger$ is applied to the vacuum state $|0\rangle$, we have

$$
\beta_p^\dagger |0\rangle = b_p^\dagger |0\rangle, \tag{2.41}
$$

and it is expressed by only the bare phonon operator (before the transformation) in the same p mode. Therefore we mainly focus on the quasiparticle expressed by $(\hat{\alpha}_i^\dagger, \hat{\alpha}_i)$ in the following section. Note that it is valid only if the bare photon number (the expectation value of $\hat{a}_i^\dagger a_i$) is not so large that the fluctuation is more important than the bare photon number. In other words, the model we are considering is suitable for discussing the quantum nature of a few photons in an optically excited probe system.

The coherent state of phonons is not an eigenstate of the Hamiltonian, and thus the number of phonons as well as the energy are fluctuating. This fluctuation allows incident photons into the probe system to excite phonon fields. When all the phonon fields are in the vacuum at time $t = 0$, the excitation probability $P(t)$ that a photon incident on site i in the model system excites the phonon mode p at time t is given by

$$P(t) = 1 - \exp\left\{2\left(\frac{\chi_{ip}}{\Omega_p}\right)^2 (\cos \Omega_p t - 1)\right\}, \qquad (2.42)$$

where the photon-hopping term is neglected for simplicity. The excitation probability oscillates at frequency of $2\pi/\Omega_p$, and has the maximum value at $t = \pi/\Omega_p$. The frequencies of the localized phonon modes are higher than those of the delocalized ones, and the localized modes at the earlier time are excited by the incident photons.

Figure 2.6 shows the temporal evolution of the excitation probability $P_{p0}(t)$ calculated from

$$P_{p0}(t) = \left[1 - \exp\left\{2\left(\frac{\chi_{ip0}}{\Omega_{p0}}\right)^2 (\cos \Omega_{p0} t - 1)\right\}\right]$$
$$\times \exp\left\{\sum_{p \neq p0} 2\left(\frac{\chi_{ip}}{\Omega_p}\right)^2 (\cos \Omega_p t - 1)\right\}, \qquad (2.43)$$

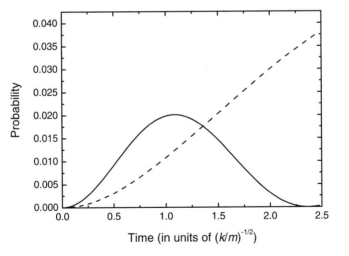

Fig. 2.6. Temporal evolution of the excitation probability of a localized (delocalized) phonon mode that is represented by the solid (dashed) curve. The system is initially excited by a photon at the impurity site 26. The coupling constant $\chi = 10.0$ (fsec^{-1} nm^{-1}) and the parameter $\hbar\sqrt{k/m} = 22.4$ meV are used, while other parameters are the same as those in Fig. 2.5

where a specific phonon mode p_0 is excited while other modes are in the vacuum state. In Fig. 2.6, the solid curve represents the probability that a localized phonon mode is excited as the p_0 mode, while the dashed curve illustrates how the lowest phonon mode is excited as the p_0 mode. It follows from the figure that the localized phonon mode is dominantly excited at the earlier time.

2.4.4 Localization Mechanism

In this section, we discuss how phonons contribute to the spatial distribution of photons in the pseudo one-dimensional system under consideration. When there are no interactions between photons and phonons, the frequency and hopping constant are equal at all sites, and thus the spatial distribution of photons are symmetric. It means that no photon localization occurs at any specific site. However, if there are any photon–phonon interactions, spatial inhomogeneity or localization of phonons affects the spatial distribution of photons. On the basis of the Hamiltonian (2.38), we analyze the contribution from the diagonal and off-diagonal parts in order to investigate the localization mechanism of photons.

2.4.4.1 Contribution from the Diagonal Part

Let us rewrite the third term of the Hamiltonian (2.38) with the mean field approximation as

$$-\sum_{i=1}^{N}\sum_{j=1}^{N}\sum_{p=1}^{N}\hbar\frac{\chi_{ip}\chi_{jp}}{\Omega_p}\hat{\alpha}_i^\dagger\hat{\alpha}_i\langle\hat{N}_j\rangle \equiv -\sum_{i=1}^{N}\hbar\omega_i\hat{\alpha}_i^\dagger\hat{\alpha}_i, \qquad (2.44)$$

with

$$\omega_i \equiv \sum_{j=1}^{N}\sum_{p=1}^{N}\frac{\chi_{ip}\chi_{jp}}{\Omega_p}\langle\hat{N}_j\rangle = \sum_{j=1}^{N}\sum_{p=1}^{N}\frac{\hbar\chi^2 P_{ip}P_{jp}}{2N\Omega_p^2(m_im_j)^{1/2}}, \qquad (2.45)$$

where (2.25) is used to obtain the expression in the last line of (2.45). In addition, we neglect the site dependence of the hopping operator $\hat{J}_i$ to approximate J, for the moment. Then the Hamiltonian regarding the quasiparticles ($\hat{\alpha}$ and $\hat{\alpha}^\dagger$) can be expressed as

$$\hat{H} = \sum_{i=1}^{N}\hbar(\omega - \omega_i)\hat{\alpha}_i^\dagger\hat{\alpha}_i + \sum_{i=1}^{N-1}\hbar J\left(\hat{\alpha}_i^\dagger\hat{\alpha}_{i+1} + \hat{\alpha}_{i+1}^\dagger\hat{\alpha}_i\right), \qquad (2.46)$$

or in matrix form as

$$\hat{H} = \hbar\hat{\alpha}^\dagger\begin{pmatrix} \omega - \omega_1 & J & & \\ J & \omega - \omega_2 & \ddots & \\ & \ddots & \ddots & J \\ & & J & \omega - \omega_N \end{pmatrix}\hat{\alpha}, \qquad (2.47a)$$

$$\hat{\alpha}^\dagger \equiv \left(\hat{\alpha}_1^\dagger, \hat{\alpha}_2^\dagger, \ldots, \hat{\alpha}_N^\dagger\right), \qquad (2.47b)$$

where the effect from the phonon fields is involved in the diagonal elements ω_i. Denoting an orthonormal matrix to diagonalize the Hamiltonian (2.47a) as Q and the rth eigenvalue as E_r, we have

$$\hat{H} = \sum_{r=1}^{N} \hbar E_r \hat{A}_r^\dagger \hat{A}_r, \tag{2.48a}$$

$$\hat{A}_r = \sum_{i=1}^{N} (Q^{-1})_{ri} \hat{\alpha}_i = \sum_{i=1}^{N} Q_{ir} \hat{\alpha}_i, \tag{2.48b}$$

$$[\hat{A}_r, \hat{A}_s^\dagger] = \delta_{rs}. \tag{2.48c}$$

Using the above relations (2.48a)–(2.48c), we can write down the time evolution of the photon number operator at site i as follows:

$$\hat{N}_i(t) = \exp\left(i\frac{\hat{H}t}{\hbar}\right) \hat{N}_i \exp\left(-i\frac{\hat{H}t}{\hbar}\right),$$

$$= \sum_{r=1}^{N} \sum_{s=1}^{N} Q_{ir} Q_{is} \exp\{i(E_r - E_s)t\} \hat{A}_r^\dagger \hat{A}_s. \tag{2.49}$$

The expectation value of the photon number operator $\hat{N}_i(t)$ is then given by

$$\langle N_i(t) \rangle_j = \langle \psi_j | \hat{N}_i(t) | \psi_j \rangle,$$

$$= \sum_{r=1}^{N} \sum_{s=1}^{N} Q_{ir} Q_{jr} Q_{is} Q_{js} \cos\{(E_r - E_s)t\}, \tag{2.50}$$

in terms of one photon state at site j defined by

$$|\psi_j\rangle = \hat{\alpha}_j^\dagger |0\rangle = \sum_{r=1}^{N} Q_{jr} \hat{A}_r^\dagger |0\rangle. \tag{2.51}$$

Since the photon number operator $\hat{N}_i$ commutes with the Hamiltonian (2.46), the total photon number is conserved, which means that a polariton called as a photon in this chapter conserves the total particle number within the lifetime. Moreover, $\langle N_i(t) \rangle_j$ can be regarded as the observation probability of a photon at an arbitrary site i and time t, initially populated at site j. This function is analytically expressed in terms of the Bessel function as

$$\langle N_i(t) \rangle_j = \{J_{j-i}(2Jt) - (-1)^i J_{j+i}(2Jt)\}^2, \tag{2.52}$$

when there are no photon–phonon interactions ($\omega_i = 0$) and the total site number N becomes infinite. Here the argument J is the photon-hopping constant, and (2.52) shows that a photon initially populated at site j delocalizes to a whole system.

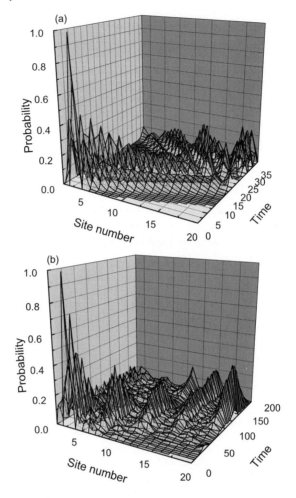

Fig. 2.7. a The probability that a photon is found at each site as a function of time in the case of $\chi = 0$ and $\hbar J = 0.5$ eV. The time scale is in units of $1/J$. Other parameters are given in the text. **b** The probability that a photon is found at each site as a function of time, in the case of $J \sim (\hbar/k)(\chi/N)^2$. The time scale is in units of $1/J$. Other parameters are given in the text

Focusing on the localized phonon modes, we take the summation in (2.45) over the localized modes only, which means that an earlier stage is considered after the incident photon excites the phonon modes, or that the duration of the localized phonon modes dominant over the delocalized modes is focused (see Fig. 2.6). This kind of analysis provides us with an interesting insight to the photon–phonon coupling constant and the photon-hopping constant, which is necessary for the understanding of the mechanism of photon's localization.

The temporal evolution of the observation probability of a photon at each site is shown in Fig. 2.7. Without the photon–phonon coupling ($\chi = 0$), a photon spreads over the whole system as a result of the photon hopping, as shown in Fig. 2.7(a).

Here the photon energy $\hbar\omega = 1.81\,\text{eV}$ and the hopping constant $\hbar J = 0.5\,\text{eV}$ are used in the calculation. The impurities are assumed to be doped at sites 3, 7, 11, 15, and 19 while the total site number N is 20 and the mass ratio of the host molecules to the impurities is 5. Figure 2.7(b) shows a result with $\chi = 1.4 \times 10^3\,\text{fsec}^{-1}\,\text{nm}^{-1}$ while other parameters used are the same as those in Fig. 2.7(a). It follows from the figure that a photon moves from one impurity to other impurity sites instead of delocalizing to a whole system. As the photon–phonon coupling constant becomes much larger than $\chi = 1.4 \times 10^3\,\text{fsec}^{-1}\,\text{nm}^{-1}$, a photon cannot move from the initial impurity site to others and stay there.

The effect due to the photon–phonon coupling χ is expressed by the diagonal component in the Hamiltonian, while the off-diagonal component involves the photon-hopping effect due to the hopping constant J. The above results indicate that the photon's spatial distribution depends on the competition between the diagonal and off-diagonal components in the Hamiltonian, i.e., χ and J, and that a photon can move among impurity sites and localize at those sites when both components are comparable under the condition

$$\chi \sim N\sqrt{\frac{k}{\hbar}}J, \qquad (2.53)$$

where the localization width seems very narrow.

2.4.4.2 Contribution from the Off-Diagonal Part

In the previous section, we have approximated J as a constant independent of the sites, in order to examine the photon's spatial distribution as well as the mechanism of the photon localization. Now let us treat the photon-hopping operator $\hat{J}_i$ more rigorously, and investigate the site dependence of the off-diagonal contribution, which includes the inhomogeneity of the phonon fields. Noticing that a quasiparticle transformed from a photon operator by the Davydov transformation is associated with phonons in the coherent state (see (2.40)), we take expectation values of $\hat{J}_i$ in terms of the coherent state of phonons $|\gamma\rangle$ as

$$J_i \equiv \langle\gamma|\hat{J}_i|\gamma\rangle. \qquad (2.54)$$

Here the coherent state $|\gamma\rangle$ is an eigenstate of the annihilation operator $\hat{b}_p$ with eigenvalue γ_p and satisfies the following equations

$$\hat{b}_p|\gamma\rangle = \gamma_p|\gamma\rangle, \qquad (2.55a)$$

$$\exp\left(-\sum_p c_p\hat{b}_p\right)|\gamma\rangle = \exp\left(-\sum_p c_p\gamma_p\right)|\gamma\rangle, \qquad (2.55b)$$

where c_p is a real number. Since the difference between the creation and annihilation operators of a phonon is invariant under the Davydov transformation, the following relation holds (see (2.36c) and (2.36d)):

$$\hat{\beta}_p^{\dagger} - \hat{\beta}_p = \hat{b}_p^{\dagger} - \hat{b}_p. \tag{2.56}$$

Using (2.39), (2.55a), (2.55b), and (2.56), we can rewrite the site-dependent hopping constant J_i in (2.54) as

$$
\begin{aligned}
J_i &= J \langle \gamma | \exp \left\{ \sum_{p=1}^{N} C_{ip} \left(\hat{b}_p^{\dagger} - \hat{b}_p \right) \right\} | \gamma \rangle \\
&= J \exp \left(-\frac{1}{2} \sum_{p=1}^{N} C_{ip}^2 \right) \langle \gamma | \exp \left(\sum_{p'=1}^{N} C_{ip'} \hat{b}_{p'}^{\dagger} - \sum_{p''=1}^{N} C_{ip''} \hat{b}_{p''} \right) | \gamma \rangle \\
&= J \exp \left(-\frac{1}{2} \sum_{p=1}^{N} C_{ip}^2 \right) \langle \gamma | \exp \left(\sum_{p'=1}^{N} C_{ip'} \gamma_{p'} - \sum_{p''=1}^{N} C_{ip''} \gamma_{p''} \right) | \gamma \rangle \\
&= J \exp \left(-\frac{1}{2} \sum_{p=1}^{N} C_{ip}^2 \right), \tag{2.57}
\end{aligned}
$$

where C_{ip} is denoted by

$$C_{ip} \equiv \frac{\chi_{i,p} - \chi_{i+1,p}}{\Omega_p}. \tag{2.58}$$

Figure 2.8 shows the site dependence of J_i in the case of $N = 20$. Impurities are doped at site 4, 6, 13 and 19. The mass ratio of the host molecules to the impurities is 5, while $\hbar J = 0.5\,\text{eV}$ and $\chi = 14.0\,\text{fsec}^{-1}\,\text{nm}^{-1}$ are used. It follows from the figure that the hopping constants are highly modified around the impurity sites and the

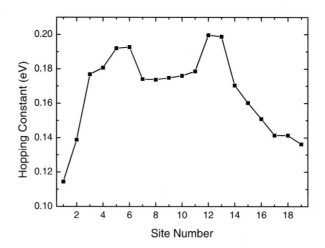

Fig. 2.8. The site dependence of the hopping constants J_i in the case of $N = 20$. Impurities are doped at sites 4, 6, 13 and 19. The mass ratio of the host molecules to the impurities is 1 to 0.2, while $\hbar J = 0.5\,\text{eV}$ and $\chi = 40.0\,\text{fsec}^{-1}\,\text{nm}^{-1}$ are used

edge sites. The result implies that photons are strongly affected by localized phonons and hop to the impurity sites to localize. Here we have not considered the temperature dependence of J_i, which is important for phenomena dominated by incoherent phonons [51]. This is because coherent phonons weakly depend on the temperature of the system. However, there remains room to discuss a more fundamental issue, i.e., whether the probe system is in a thermal equilibrium state or not.

In Fig. 2.9, we present a typical result, that photons localize around the impurity sites in the system as the photon–phonon coupling constants χ vary from zero to $40.0 \, \mathrm{fsec}^{-1} \, \mathrm{nm}^{-1}$ or $54.0 \, \mathrm{fsec}^{-1} \, \mathrm{nm}^{-1}$ while keeping $\hbar J = 0.5 \, \mathrm{eV}$. As depicted with the filled squares in the figure, photons delocalize and spread over the system without the photon–phonon couplings. When the photon–phonon couplings are comparable to the hopping constants, $\chi = 40.0 \, \mathrm{fsec}^{-1} \, \mathrm{nm}^{-1}$, photons can localize around the impurity site with a finite width, two sites at HWHM, as shown with the filled circles. This finite width of photon localization comes from the site-dependent hopping constants. As the photon–phonon couplings are larger than $\chi = 40.0 \, \mathrm{fsec}^{-1} \, \mathrm{nm}^{-1}$, photons can localize at the edge sites with a finite width, as well as the impurity sites. In Fig. 2.9, the photon localization at the edge site is shown with the filled triangles, which originates from the finite size effect of the molecular chain [43, 52]. This kind of localization of photons, dressed by the coherent state of phonons, leads us to a simple understanding of phonon-assisted photodissociation using an optical near field. Molecules in the electronic ground state approach the probe tip within the localization range of the dressed photons, and can be vibrationally excited by the dressed-photon transfer to the molecules. This occurs via the multiphonon component of the dressed photons, which might be followed by the electronic excitation.

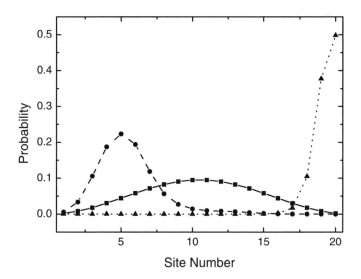

Fig. 2.9. Probability of photons observed at each site. The filled squares, circles and triangles represent the results for $\chi = 0$, 40.0 and $54.0 \, \mathrm{fsec}^{-1} \, \mathrm{nm}^{-1}$, respectively. Other parameters are the same as those in Fig. 2.8

Thus it leads to the dissociation of the molecules even if the incident photon energy is less than the dissociation energy used.

2.5 Conclusions

As a natural extension of the localized photon model, we discussed the inclusion of phonons' effects into the model. The study was initially motivated by experiments on photodissociation of molecules by optical near fields. Those results show unique features, different from the conventional results with far fields. After clarifying delocalized or localized vibration modes in a pseudo one-dimensional system, we focused on the interaction between dressed photons and phonons using the Davydov transformation. We theoretically showed that photons are dressed by the coherent state of phonons, and found that the competition between the photon–phonon coupling constant and the photon-hopping constant governs the photon localization or delocalization in space. The results lead us to a simple understanding of an optical near field itself as an interacting system of photon, electronic excitation (induced polarization) and phonon fields in a nanometer space, which are surrounded by macroscopic environments, as well as phonon-assisted photodissociation using an optical near field.

Acknowledgments

The authors are grateful to H. Hori (Yamanashi Univ.), S. Sangu (Ricoh Co., Ltd.), A. Shojiguchi (NEC Co.), K. Kitahara (International Christian Univ.), T. Yatsui (Japan Science and Technology Agency), M. Tsukada (Waseda Univ.), H. Nejo (National Institute for Materials Science), M. Naruse (National Institute of Information and Communications Technology), M. Ikezawa (Univ. of Tsukuba), A. Sato (Tokyo Institute of Technology), H. Ishihara (Osaka Prefecture Univ.) and I. Banno (Yamanashi Univ.) for stimulating discussions. This work was supported in part by the 21st Century COE program at Tokyo Institute of Technology "Nanometer-Scale Quantum Physics" and by a Grant-in-Aid for Scientific Research from the Ministry of Education, Culture, Sports, Science and Technology, Japan, and by CREST, Japan Science and Technology Agency.

References

[1] M. Ohtsu, K. Kobayashi, *Optical Near Fields* (Springer, Berlin, 2004)
[2] N. Hosaka, T. Saiki, J. Microsc. **202**, 362 (2001)
[3] K. Matsuda, T. Saiki, S. Nomura, M. Mihara, Y. Aoyagi, S. Nair, T. Takagahara, Phys. Rev. Lett. **91**, 177401 (2003)
[4] Y. Yamamoto, M. Kourogi, M. Ohtsu, V. Polonski, G.H. Lee, Appl. Phys. Lett. **76**, 2173 (2000)
[5] T. Kawazoe, Y. Yamamoto, M. Ohtsu, Appl. Phys. Lett. **79**, 1184 (2001)

[6] T. Yatsui, M. Ueda, Y. Yamamoto, T. Kawazoe, M. Kourogi, M. Ohtsu, Appl. Phys. Lett. **81**, 3651 (2002)

[7] T. Kawazoe, K. Kobayashi, S. Sangu, M. Ohtsu, Appl. Phys. Lett. **82**, 2957 (2003)

[8] K. Kobayashi, S. Sangu, A. Shojiguchi, T. Kawazoe, K. Kitahara, M. Ohtsu, J. Microsc. B **210**, 247 (2003)

[9] S. Sangu, K. Kobayashi, A. Shojiguchi, M. Ohtsu, Phys. Rev. B **69**, 115334 (2004)

[10] T. Kawazoe, K. Kobayashi, M. Ohtsu, Appl. Phys. Lett. **86**, 103102 (2005)

[11] I. Bialynicki-Birula, Photon wave function, in *Progress in Optics*, vol. 36, ed. by E. Wolf (North-Holland, Amsterdam, 1996), pp. 248–294

[12] K. Cho, *Optical Response of Nanostructures: Microscopic Nonlocal Theory* (Springer, Berlin, 2003)

[13] K. Kobayashi, S. Sangu, M. Ohtsu, Quantum theoretical approach to optical near-fields and some related application, in *Progress in Nano-Electro-Optics I*, ed. by M. Ohtsu (Springer, Berlin, 2003), pp. 119–157

[14] O. Keller, Phys. Rep. **411**, 1 (2005)

[15] A. Shojiguchi, K. Kobayashi, S. Sangu, K. Kitahara, M. Ohtsu, J. Phys. Soc. Jpn. **72**, 2984 (2003)

[16] A. Shojiguchi, K. Kobayashi, S. Sangu, K. Kitahara, M. Ohtsu, A phenomenological description of optical near fields and optical properties of *N* two-level systems inter-acting with optical near fields, in *Progress in Nano-Electro-Optics III*, ed. by M. Ohtsu (Springer, Berlin, 2005), pp. 145–220

[17] T. Kawazoe, K. Kobayashi, J. Lim, Y. Narita, M. Ohtsu, Phys. Rev. Lett. **88**, 067404 (2002)

[18] K. Kobayashi, S. Sangu, T. Kawazoe, M. Ohtsu, J. Lumin. **112**, 117 (2005); **114**, 315 (2005)

[19] M. Scheibner, T. Schmidt, L. Worschech, A. Forchel, G. Bacher, T. Passow, D. Hommel, Nature Phys. **3**, 106 (2007)

[20] T. Kawazoe, K. Kobayashi, S. Takubo, M. Ohtsu, J. Chem. Phys. **122**, 024715 (2005)

[21] K. Kobayashi, T. Kawazoe, M. Ohtsu, IEEE Trans. Nanotech. **4**, 517 (2005)

[22] T. Kawazoe, K. Kobayashi, M. Ohtsu, Appl. Phys. B **84**, 247 (2006)

[23] J.J. Sakurai, *Advanced Quantum Mechanics* (Addison-Wesley, Reading, 1967)

[24] T.D. Newton, E.P. Wigner, Rev. Mod. Phys. **21**, 400 (1949)

[25] J.E. Sipe, Phys. Rev. A **52**, 1875 (1995)

[26] M.O. Scully, M.S. Zubairy, *Quantum Optics* (Cambridge University Press, Cambridge, 1997)

[27] K. Kobayashi, M. Ohtsu, J. Microsc. **194**, 249 (1999)

[28] K. Kobayashi, S. Sangu, H. Ito, M. Ohtsu, Phys. Rev. A **63**, 013806 (2001)

[29] S. John, T. Quang, Phys. Rev. A **52**, 4083 (1995)

[30] H. Suzuura, T. Tsujikawa, T. Tokihiro, Phys. Rev. B **53**, 1294 (1996)

[31] C.R. Kagan, C.B. Murray, M. Nirmal, M.G. Bawendi, Phys. Rev. Lett. **76**, 1517 (1996)

[32] S.A. Crooker, J.A. Hollingsworth, S. Tretiak, V.I. Klimov, Phys. Rev. Lett. **89**, 186802 (2002)

[33] T. Yatsui, M. Ohtsu, S.J. An, J. Yoo, G.-C. Yi, Appl. Phys. Lett. **87**, 033101 (2005)

[34] G. McDermott, S.M. Prince, A.A. Freer, A.M. Hawthornthwaite-Lawless, M.Z. Papiz, R.J. Cogdell, N.W. Isaacs, Nature **374**, 517 (1995)

[35] K. Mukai, S. Abe, H. Sumi, J. Phys. Chem. B **103**, 6096 (1999)

[36] R.L. Jackson, Chem. Phys. Lett. **163**, 315 (1989)

[37] R.L. Jackson, J. Chem. Phys. **96**, 5938 (1992)

[38] R. Schinke, *Photodissociation Dynamics* (Cambridge University Press, Cambridge, 1993)

[39] H. Haken, H.C. Wolf, *Molecular Physics and Elements of Quantum Chemistry* (Springer, Berlin, 1995)
[40] T. Kawazoe, K. Kobayashi, M. Ohtsu, Appl. Phys. B **84**, 247 (2006)
[41] J.J. Hopfield, Phys. Rev. **112**, 1555 (1958)
[42] A.L. Ivanov, H. Haug, L.V. Keldysh, Phys. Rep. **296**, 237 (1998)
[43] C. Falvo, V. Pouthier, J. Chem. Phys. **122**, 014701 (2005)
[44] M.E. Striefler, G.R. Barsch, Phys. Rev. B **12**, 4553 (1975)
[45] D. Paton, W.M. Visscher, Phys. Rev. **154**, 802 (1967)
[46] A.J. Sievers, A.A. Maradudin, S.S. Jaswal, Phys. Rev. **138**, A272 (1965)
[47] S. Mizuno, Phys. Rev. B **65**, 193302 (2002)
[48] T. Yamamoto, K. Watanabe, Phys. Rev. Lett. **96**, 255503 (2006)
[49] A.S. Davydov, G.M. Pestryakov, Phys. Stat. Sol. B **49**, 505 (1972)
[50] L. Jacak, P. Machnikowski, J. Krasnyj, P. Zoller, Eur. Phys. J. D **22**, 319 (2003)
[51] K. Mizoguchi, T. Furuihi, O. Kojima, M. Nakayama, S. Saito, A. Syouji, K. Sakai, Appl. Phys. Lett. **87**, 093102 (2005)
[52] V. Pouthier, C. Girardet, J. Chem. Phys. **112**, 5100 (2000)

3

Visible Laser Desorption/Ionization Mass Spectrometry Using Gold Nanostructure

L.C. Chen, H. Hori, and K. Hiraoka

3.1 Introduction

When the nanostructured surface of a highly conductive metal, e.g. gold, is irradiated with a laser of certain wavelength at appropriate polarization, collective electron motion, known as localized surface plasmon–polariton oscillation will be excited. The localized surface plasmon–polariton resonance leads to enhanced photon absorption and huge concentration of optical near-fields at a small volume, which contribute to the enhancement in the surface-enhanced spectroscopy. Although intensive research on the plasmonic electronics and plasmon biosensing is in progress, there is little work on the exploitation of the plasmon effect in the desorption/ionization of biomolecules for mass spectrometry.

In this report, we describe the visible laser desorption/ionization of biomolecules from the gold-coated porous silicon, gold nanorod arrays, and nanoparticles. The porous silicon made by electrochemical etching was coated with gold using argon ion sputtering. The gold nanorod arrays were fabricated by electro-depositing the gold into the porous alumina template, and the subsequent partial removal of the alumina template. A frequency-doubled Nd:YAG laser was used to irradiate gold nanostructured substrate, and the desorbed molecular ions were mass analysed by a time-of-flight mass spectrometer (TOF-MS). The present technique offers a potential analytical method for the low-molecular weight analytes which are rather difficult to handle in the conventional matrix-assisted laser desorption/ionization (MALDI) mass spectrometry. With the presence of Au nanoparticles, the UV-MALDI matrix was also found to be photo-ionized by the 532-nm laser even though the photon energy is insufficient for free molecules.

3.1.1 Matrix-Assisted Laser Desorption/Ionization Mass Spectrometry

Mass spectrometry is a very powerful analytical tool for biochemistry, pharmacy and medicine. The basic principle of mass spectrometry is to generate ions from the inorganic or organic compound, and to separate these ions according to their mass-to-charge ratio (m/z). The analyte may be ionized by a variety of methods, for example, electron ionization [1], electrospray [2], and laser desorption/ionization.

For laser mass spectrometry, matrix-assisted laser desorption/ionization (MALDI) is a very effective and soft method in obtaining mass spectra for synthetic and biological samples, such as peptides and proteins with less molecular fragmentation [3, 4]. Depending on the matrices, laser wavelengths of ultraviolet (UV) and infrared have been employed. The established UV-MALDI method usually employs a nitrogen laser (337 nm), or a frequency-tripled Nd:YAG laser (355 nm) for desorption/ionization, while the Er:YAG lasers (2.94 μm) and CO_2 (10.6 μm) are used in the IR-MALDI.

In the MALDI, the biomolecular analytes are mixed with the photo-absorbing chemical matrix of suitable functional groups to assist the energy transfer. Thus, the matrix molecules must possess suitable chromophores to absorb the laser photons efficiently. Nicotinic acid (NA) was historically the first UV-MALDI matrix for successful detection of peptides and proteins. Ever since, many other better matrices, e.g., 2,5-dihydroxybenzoic acid (DHB), and α-cyano-4-hydroxycinnamic acid (CHCA) have also been found. As for IR-MALDI, the laser wavelength of $\sim$3 μm effectively excites the O–H and N–H stretch vibrations of the molecules, while the laser wavelength of $\sim$10 μm causes the excitation of C–O stretch and O–H bending vibrations [5, 6].

Except for the light-absorbing analytes, direct photoionization of macromolecules rarely takes place and the peptides and proteins ions observed in laser desorption/ionization are mostly protonated (molecular ions generated by proton attachment, e.g., $[M + H]^+$). For the analytes such as underivatized carbohydrates, due to their poor proton affinity, the molecules are difficult to be protonated and instead are mostly ionized by alkali metals attachment, e.g., $[M + Na]^+$, $[M + K]^+$ [7]. Typical ion species produced by LDI/MALDI are listed in Table 3.1.

The typical laser fluences in UV-MALDI are in the range of 10–100 mJ cm^{-1}, which correspond to 10^6–10^7 W cm^{-2} for a pulsed laser of 10 ns pulse width. For

Table 3.1. Ions produced by LDI/MALDI

Ion species	Positive ions	Negative ions
Radical	$M^{+\bullet}$	$M^{-\bullet}$
Protonated/deprotonated	$[M + H]^+$	$[M - H]^+$
Alkali adducts	$[M + Na]^+$, $[M + K]^+$, etc.	
Cluster	$n[M + H]^+$	$n[M - H]^-$
Multiple charged	$[M + 2H]^{2+}$,	$[M - 2H]^{2-}$
	$[M + 2Na]^{2+}$, etc.	

IR-MALDI, the required fluence is 10 times higher than that of UV-MALDI. At the laser threshold, a sharp onset of desorption/ionization takes place. The threshold laser fluence depends on the type of matrix as well as the matrix-to-analyte mixing ratio. The mixing ratio of 5000 is usually used in MALDI. Although the detailed desorption/ionization mechanism of MALDI is not thoroughly understood, it is generally believed that the radical ions of the matrix molecules produced by either two photon ionization or the molecular exciton pooling, play crucial roles in ionizing the desorbed analytes via gas phase interaction [8, 9].

Despite efforts by various research groups [10, 11], a suitable chemical matrix for visible laser has not been found thus far, and the existing matrices are not accessible by the laser wavelength ranging from 400 nm to ∼2.7 µm [12].

3.1.2 Laser Desorption/Ionization with Inorganic Matrix and Nanostructure

Although MALDI is highly sensitive for the large biomolecules (>700 Da), the detection for the analytes of low molecular weight is rather difficult due to the matrix interferences. Thus in the low mass range, direct laser desorption/ionization (LDI) on surface modified substrates, or the use of inorganic matrices becomes the alternative to chemical MALDI. The use of nanoparticles as efficient UV-absorbing matrices was first introduced by Tanaka et al. [3], of which 30-nm cobalt powders were suspended in glycerol solution. A variety of nanomaterials—for example, titanium nitride [12], zinc oxide and titanium oxide [13] and gold nanoparticles [14]—have been proposed as inorganic matrices.

Direct LDI on various substrates—for example, graphite [15], silicon, titania sol-gel [16] and metal coated porous alumina [17]—have also been studied. In particular, porous silicon, which has high absorption in the ultraviolet region, has received considerable attention due to its reported high sensitivity [18, 19]. This method is called desorption/ionization on silicon (DIOS). Other silicon-based substrates include silicon nanowires [20], column/void silicon network [21], nanogrooves [17], and nanocavities [22]. Matrix-less IR laser desorption/ionization has also been reported on a flat silicon surface [23].

3.1.3 Time-of-Flight Mass Spectrometry

Molecular ions can be analyzed by a number of different instruments, e.g., time-of-flight mass spectrometer (TOF-MS), magnetic sector mass spectrometer, quadrupole ion trap mass spectrometer (QIT-MS), and Fourier transform ion cyclotron resonance mass spectrometer (FT-ICR-MS) [5, 6]. In particular, the TOF-MS is used almost exclusively for ions produced by laser desorption/ionization. The ions generated upon the irradiation of a pulsed laser are separated during their flight along the field-free path, and arrive at the detector at different times depending on their m/z. The flight time is given as:

$$t = \frac{L}{\sqrt{2eU}} \sqrt{\frac{m}{z}}, \qquad (3.1)$$

where e the electron charge, L is the flight tube length and U is the accelerating voltage. Thus, m/z is proportional to t^2.

For TOF-MS, the initial ion velocity produced from the laser plume, and the initial position of the generated ions can contribute to measurement errors and reduced resolution of the instrument. To a certain extent, the initial velocities effect can be reduced by using the delayed extraction, in which the accelerating voltage is applied only after a certain time interval (typically tens to hundreds of ns) after the laser pulse [5]. The initial velocity/energy distribution can also be compensated for by using an ion reflector or a reflectron.

3.2 Surface Plasmon–Polariton

Surface plasmon is a collective oscillation of electron density on the metal surface [24]. At surface plasmon resonance, all the free electrons within the conduction band oscillate in phase and lead to a huge concentration of the electric field at a small volume (Fig. 3.1). The plasmonic oscillation which is coupled with the optical field is usually referred to as the plasmon–polariton.

For the nanoparticles of noble metals, e.g., silver and gold, the surface plasmon–polariton resonance takes place in visible light, and their optical properties can be described by the Mie-extinction, σ_{ext}, and

$$\sigma_{ext} = \sigma_{abs} + \sigma_{sca}, \tag{3.2}$$

where σ_{abs}, and σ_{sca} are the absorption and scattering cross-section, respectively. Figure 3.2 shows the calculated normalized extinction cross-section of the nanoparticles for various noble metals. The particles' diameters are assumed to be 40 nm, and the Mie-extinction [25, 26] is solved numerically. For gold, silver and copper,

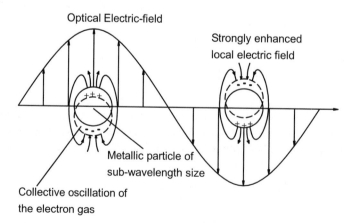

Fig. 3.1. The plasmon–polariton oscillation of metallic particles which have diameters much smaller than the wavelength under the illumination of polarized optical wave

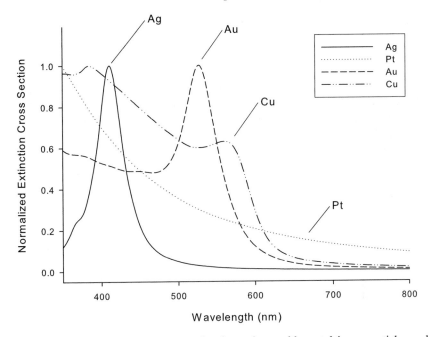

Fig. 3.2. Normalized extinction cross-section for various noble metals' nanoparticles: gold (Au), silver (Ag), copper (Cu) and platinum (Pt)

the nanoparticles have unique absorption bands in the visible region (see Fig. 3.2), and they can be easily identified by the colour of their scattered light. The intense red colour of the aqueous dispersion of the colloidal gold nanoparticles is the manifestation of the localized surface plasmon–polariton resonance, which peak at ∼520 nm.

If the nanoparticles are much smaller than the exciting optical wavelength ($r \ll \lambda$, where r is the radius of the nanoparticle and λ is the optical wavelength), the extinction cross-section is primarily due to the dipole oscillation, and the Mie theory reduces to dipole approximation [27, 28]:

$$\sigma_{\text{ext}}(\omega) = 9\frac{\omega}{c}\varepsilon_m^{3/2}V \frac{\varepsilon_2(\omega)}{[\varepsilon_1(\omega) + 2\varepsilon_m]^2 + \varepsilon_2(\omega)^2}, \tag{3.3}$$

where $V = \frac{4}{3}\pi r^3$, ω is the angular frequency of the exciting light, c is the speed of light, $\varepsilon(\omega) = \varepsilon_1(\omega) + i\varepsilon_2(\omega)$ is the dielectric function of the nanoparticles, and ε_m is dielectric function of the surrounding medium. The resonance condition for surface plasmon–polariton is fulfilled when

$$\varepsilon_1(\omega) = -2\varepsilon_m \tag{3.4}$$

if $\varepsilon_2(\omega)$ is small or weakly dependent on ω [27].

The surface plasmon-induced electromagnetic field enhancement on the metallic nanoparticles had been known to be accountable for the surface-enhanced Raman spectroscopy (SERS) [29], as well as nonlinear optical responses such as second harmonic generation [30], and optical frequency mixing [31]. Besides electromagnetic enhancement, the metal-absorbates electronic coupling had also been known to contribute to the chemical enhancement for SERS [32, 33]. Upon absorption on the metal surface, the interaction between the absorbate molecules and the electron gas on the metal surface results in the broadening and shifting in energy of the free molecular states [32]. Thus, even though the states' transition of the free molecule may be too energetic to be excited by say, a visible laser, a near resonance could be found for the laser once the molecule is adsorbed on the metal surface.

The exploitation of localized surface plasmon–polariton resonance using ordered nanoparticles arrays and aggregates include optical near-field lithography [34], and plasmonic waveguides [35]. Functionalized or conjugated gold nanoparticles, which have high binding affinity to specific analytes, are used for biosensing [36] and DNA detection [37]. Selective laser photothermal therapy using nanoparticles has also been proposed for cancer treatment [38].

3.2.1 Plasmon-Induced Desorption

3.2.1.1 Desorption of Metallic Ions

Because the surface plasmon resonance is strongly damped, the local heating—due to the joule losses on the metal surface—could take place. For the gold nanoparticles which have small heat capacity, the heat transfer was estimated to be in the picosecond time scale, and the high lattice temperature can be reached rapidly [28]. However, due to the strong electron oscillation, the plasmon induced nonthermal desorption had been reported for several metals. By irradiating the roughened silver surface with a visible or near-ultraviolet laser, two prominent peaks were observed in the kinetic-energy distribution of Ag^+ ions produced from the surface [39].

In another experiment where the surface plasmon was coupled using the attenuated-total-reflection method, similar results were also obtained for the metal atoms desorbed from the Au, Al and Ag films [40]. Although the lower energy peak was generally referred to as the thermal peak, they attributed the peak of the higher kinetic energy to the nonthermal electronic process. Nonthermal visible laser desorption of alkali atoms was also reported for sodium particles and sodium film [41, 42]. A theoretical model that involves energy coupling of surface plasmon via ion collision has also been put forward to support the plasmon hypothesis [43].

Desorption of Absorbates

On the metal surface, the adsorbed molecules produce physical or chemical bonding via, at least in part, interaction with the electron gas on the metal substrate. At plasmon resonance, due to the collective motion of electron gas, the strong optical near-field enhancement and associated strong modulation of electronic energy levels take place (Fig. 3.3).

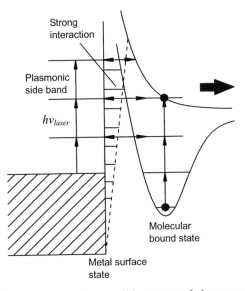

Fig. 3.3. Schematic diagram showing the possible process of plasmon-assisted desorption of adsorbates

This plasmonic process near the metallic surface produces the highly excited plasmonic sideband states due to the nonlinear optical effect. Regardless of the nature of the bonding between molecules and metallic substrate, the plasmonic sideband formation results in the optical near-field excitation of the bonding, and some of the highly excited bonding states can exert stochastic transition into the dissociation states of the absorbed molecules (see Fig. 3.3).

3.3 Visible Laser Desorption/Ionization on Gold Nanostructure

Although intensive research on the plasmonic electronics and plasmon biosensing is in progress [44, 45], there is little work on the exploitation of the plasmon effect in the desorption/ionization of biomolecules for mass spectrometry. Recently, we demonstrated the use of gold nanostructure for a nonorganic matrix-based laser desorption/ionization [46, 47].

Two different substrates were tested in our experiments: gold-coated porous silicon, and gold nanorod arrays. The porous silicon with random structure was used as the nanostructured template, and was coated with gold using argon ion sputtering. Depending on the type of the silicon, the nanostructure of the porous silicon can be tailored using the etching condition [48, 49].

The vertically aligned gold nanorod arrays, which had more regular surface morphology, were fabricated by electro-deposition of gold into the nanopores of the porous alumina template [50, 51]. The diameter of the gold nanorods follows the pores of the alumina template and the aspect ratio can be controlled through the deposition time. The porous alumina template with ordered nanopore arrays can be easily fabricated using anodic oxidation. The pore diameter can be tuned from ~10 nm–100 nm depending on the electrolyte and the anodization voltage [52, 53].

3.3.1 Fabrication of Gold-Coated Porous Silicon

Owing to its photoluminescence properties, porous silicon has attracted considerable research interest since its first discovery by Canham [48]. Porous silicon can be fabricated easily using electrochemical etching. Depending on the type of silicon, etching parameters such as etching current, time, etchant concentration and illumination are reported to affect the pore size and the porosity of the etched silicon [49].

Owing to its high UV absorption, porous silicon has also been used as substrate in direct UV-LDI (DIOS). Encouraged by the success of DIOS, the morphology of porous silicon with random structure was used as the fabrication template, and was coated with gold using argon ion sputtering. Note that in our experiment, however, the operating laser wavelength was different from that of DIOS.

Anodic Etching with Hydrofluoric Acid

The porous silicon in our experiment was made by anodic etching of $0.02\,\Omega\,cm$ n-type silicon (Nilaco, Japan) using aqueous solution of ~23 wt.% hydrofluoric acid (HF). The etching was conducted at ~5 mA/cm^2 for 2 min under white light LED illumination. The etching was performed in a Teflon etching cell with platinum as the counter electrode. A super bright LED produced approximately 5 mW/cm^2 front illumination to the etching surface. Schematics in Fig. 3.4 show the electrochemical etching of the silicon using a Teflon etching cell. With illumination, macro pores

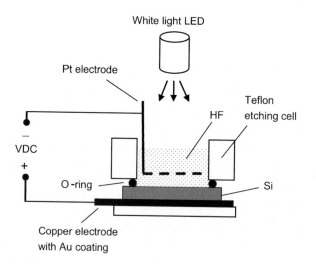

Fig. 3.4. Electrochemical etching of the silicon using Teflon etching cell

with diameter of 50 nm to 100 nm and pore depth of 100 to 200 nm were formed (microporous: dia. < 2 nm, mesoporous: dia. $= 2$–50 nm, macroporous: dia. > 50 nm).

Post-etching

When the freshly etched porous silicon was coated with gold using argon ion sputtering coater, the pores appeared to be fully covered by the gold and lost its nanostructural identity. This substrate produced no observable ion signal for most of the analytes for 532-nm laser irradiation. To increase the pore size, the freshly etched porous silicon was further treated with piranha ($H_2SO_4/H_2O_2 = 1/3$) for 4 min, followed by 10% HF etching. The piranha treatment oxidized the porous silicon lightly and formed a thin layer of silicon oxide. After stripped by 10% HF, the pores were enlarged to 100–200 nm.

Metalization of Porous Silicon

The porous silicon was metalized with gold using argon ion sputtering coater with thickness control. Inspection using scanning electron microscopy showed that the coating was not a continuous layer, and gold formed particles on the porous silicon structure with its size about the thickness of the coating. The porous silicon with and without gold coating are shown in Fig. 3.5. The cross-sectional view of the gold-coated porous silicon is depicted in Fig. 3.6. The depths of the irregular pore range are in the range of few hundreds nm.

The coated surface was also analyzed using Auger electron spectroscopy to confirm the complete coverage of gold on the porous silicon structure. The specular reflectivity of the gold-coated porous silicon is shown in Fig. 3.7. The measurement was made at normal incidence. The macroporous silicon is an efficient light trap, and it has high optical extinction extending to the visible region. After being coated

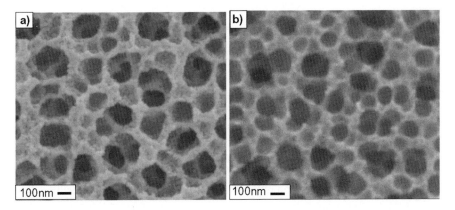

Fig. 3.5. Porous silicon **a** with ∼15-nm gold coating, and **b** without coating

Fig. 3.6. SEM image showing the cross-sectional view of the gold-coated porous silicon. The depths of the irregular pore range are in the range of a few hundreds of nm.

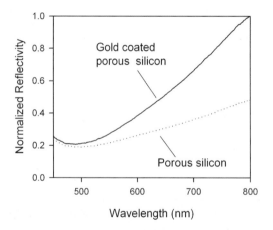

Fig. 3.7. Normalized reflectivity of porous silicon with and without gold coating

with ∼10 nm gold, its reflectivity increased significantly in the red and near-infrared region. However, the visible region of ∼500 nm remained very much unchanged due to the surface plasmon–polariton resonance.

Throughout the experiment, the porous silicon was coated with 10–15 nm thick gold. The porous silicon of other different etching conditions had also been examined for their performance in desorption/ionization. Decreasing the etching current and the strength of light illumination reduced the pores' density, size and depth, and the ion signals became weaker. Increasing the pore depth by longer etching time and current did not improve the ion yields.

3.3.2 Gold Nanorod Arrays

In order to better understand the desorption/ionization from the metallic nanostructure, it is desirable to have substrates with more regular and better-defined surface morphology. The lithography methods offer the best control over the nanostructure size, shape and spacing, but the techniques are expensive and with limited effective area. In comparison, template methods are inexpensive and can be used to pattern a large area of surface.

Nanoporous anodic alumina has been a favorite template material or mask for fabricating nanoparticle arrays [54, 55]. With suitable electrolytes and appropriate anodization condition, a high density of ordered pores can be easily formed. The pore diameter can be tuned from $\sim$10 nm to $>$100 nm by varying the anodization condition [52–54, 56]. Usually sulfuric acid is used for fabricating the alumina of pore size 10–20 nm, whereas oxalic acid, and phosphoric acid are used for bigger pore size. The pore size generally increases with the anodization voltage, however, the self-ordering takes place only under limited voltage conditions.

Porous alumina membranes were first used by Martin et al. to synthesize gold nanorods [57, 58]. The Au was electrochemically deposited within the pores and, subsequently, the Au nanorods were released and redispersed into organic solvent, followed by polymer stabilization. Because the excessive use of organic chemicals and polymers (which are essential to stabilize the nanoparticles), would likely contribute to the background noise in the mass spectrum, a modified approach was adopted in this work.

Fabrication of Nanorod Arrays

In our experiment, the embedded Au nanorods were partially released, and then held by the template, preventing the aggregation of particles without using a stabilizing agent. A schematic describing the fabrication processes is depicted in Fig. 3.8.

Aluminum sheet or the aluminum film coated on the glass or silicon substrate was used as the starting material. Sulfuric acid ($\sim$20 wt.%) was used as the electrolyte for the anodic oxidation of aluminum. A platinum counter electrode was used in the anodic oxidation as well as in the electro-deposition of the gold. The aluminum was oxidized at the anodization voltage of $\sim$12 V for 5–10 min to form porous alumina. The pore diameter was in the range of $\sim$15 nm. As shown in Fig. 3.8(a), a thin barrier layer was also formed at the bottom of the pores. Although it was possible to remove the barrier layer by etching, at $\sim$12 V anodization voltage, the barrier was thin enough that the gold could be electro-deposited directly within the pores at moderate voltage. The aqueous solution of 40 mM chlorauric acid (HAuCl$_4$) was used as the working electrolyte. The pulsed electro-deposition was conducted at $\sim$12 V with the duty cycle of 1/10 and pulse repetition rate of 1 s^{-1}. After several minutes, the deposited surface became ruby red in color, and the grown nanorods were embedded inside the porous alumina as illustrated in Fig. 3.8(b).

We note that the quality of the alumina template also depends on its initial surface roughness, and that multiple anodizing steps are usually used to produce high-quality alumina templates [55]. Occasionally, we oxidize the aluminum at higher

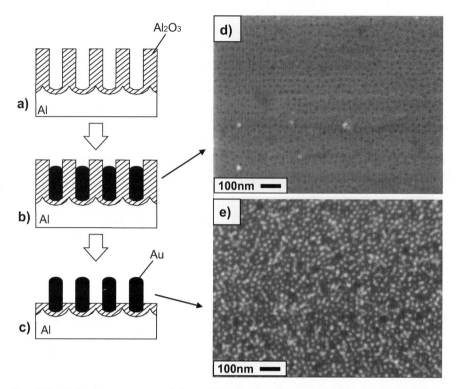

Fig. 3.8. Fabrication processes of the gold nanorods substrate. **a** Porous alumina (Al_2O_3) template fabricated by anodic oxidation using aqueous solution of $\sim$ 20 wt.% sulfuric acid. **b** Pulsed electro-deposition of gold within the pores of the porous alumina. **c** Partial removal of the alumina template using an aqueous solution of 8%v/v phosphoric acid. **d** The scanning electron micrograph of the porous alumina template embedded with gold nanorods. **e** The nanorods emerged after partial removal of the alumina template using an aqueous solution of phosphoric acid

voltage (e.g., $\sim$20 V) for a few minutes, and gradually reduce the anodization voltage to $\sim$12 V. Higher anodization voltage is known to produce more ordered and larger nanopores [52, 53, 56]. Beginning the anodization with slightly higher voltage has been found to result in more homogeneous deposition of gold onto the template.

Post-etching of the Alumina Template

To trap the analyte molecules on the gold surface, the embedded nanorods were partially exposed to the surface (Fig. 3.8(c)) by chemical etching of the alumina template. The etching was done using an aqueous solution of $\sim$8%v/v phosphoric acid.

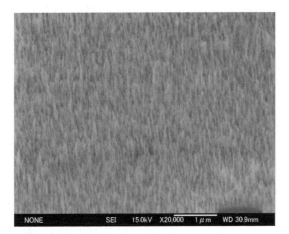

Fig. 3.9. SEM image of the gold nanorods viewed at a 45° tilt angle

Figures 3.8(d) and 3.8(e) show the SEM micrographs of the porous alumina embedded with gold nanorods, and the appearance of gold nanorods after partial removal of the alumina template, respectively. The diameter of the gold nanorods was ~15 nm and the average lengths could be fabricated in the range of ~50 to ~200 nm depending on the deposition condition. The SEM image of the fabricated gold nanorods viewed at a 45° tilt angle is shown in Fig. 3.9.

All the rods were oriented in the same direction with their major (long) axis perpendicular to the surface. Thus, the oscillation direction of the localized plasmon resonance could be selectively excited by the TM, or by TE polarized light. The optical electric field was perpendicular to the substrate surface (i.e., along the major axis of the gold nanorods) when it was TM polarized, and vice versa when it was TE polarized. Unless otherwise stated, the standard substrate used in this study consisted of gold nanorods with length of ~100 nm and diameter of ~15 nm.

3.3.2.1 Reflectivity of the Gold Nanorods

Figure 3.10 shows the normalized reflectivity of the gold nanorod substrate measured at normal incidence (light source not polarized). Compared to the porous silicon, the substrate consisted of vertically aligned gold nanorod arrays that possess a more regular surface morphology. Owing to its ordered structure, the reflectivity of the gold nanorod substrate shows a distinct optical absorption at ~520 nm, which coincides spectrally with the surface plasmon resonance of spherical nanoparticles. These visible absorption bands can be excited efficiently by a frequency doubled Nd:YAG laser at 532 nm.

Figure 3.10(b) shows the specular reflectivity of the gold nanorod substrate measured using the 532-nm laser with different optical polarization. The length of the nanorods was ~100 nm, and the measurement was taken at a 60° incidence angle, which was close to our LDI experimental condition. Normalized reflectivity of flat

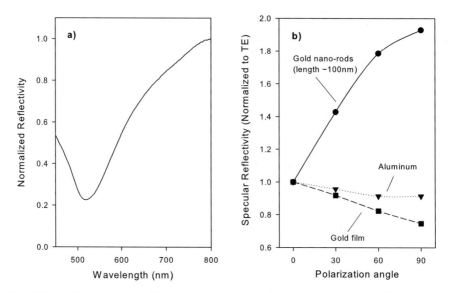

Fig. 3.10. a Normalized reflectivity of gold nanorods. **b** Specular reflectance of gold nano-rods (●), aluminum (▼), and gold film (■), measured at a 60° incidence angle using a 532-nm laser with different polarization angle. The reflected intensity is normalized to that of TE-polarization

aluminum and gold film are included for comparison. The reflected optical intensity was normalized to that of TE polarization. Unlike the flat metal in which the reflec-tivity is minimum for TM-polarized light [59], the ~100 nm gold nanorods have a higher optical absorption for the 532-nm laser at TE-polarization due to the trans-verse plasmon resonance [28, 51].

3.4 Experimental Details

3.4.1 Time-of-Flight Mass Spectrometer

The laser desorption/ionization experiment was performed with a 2.5-meter time-of-flight mass spectrometer (JEOL 2500) with delayed ion extraction. The instrument can be operated in linear or reflectron mode. A simplified schematic showing the mass spectrometer in reflectron mode is shown in Fig. 3.11. The acceleration volt-age for ions was 20 kV. The vacuum pressures in the ion source and the detector were 7.5×10^{-5} and 5×10^{-7} torr, respectively. The primary laser source for the desorption/ionization experiment was a frequency doubled Nd:YAG laser which was operated at 532-nm wavelength and pulse width of 4 ns. The gold nanostructured substrate was attached to a modified target plate and was irradiated by the laser at 60° to the surface normal. Pictures showing the time-of-flight spectrometer and the target plated are depicted in Fig. 3.12 and 3.13, respectively.

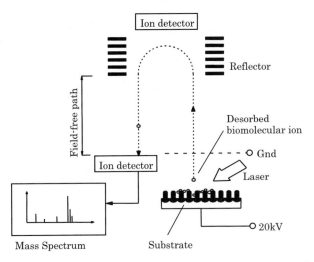

Fig. 3.11. Simplified schematic of the time-of-flight mass spectrometer in reflectron mode

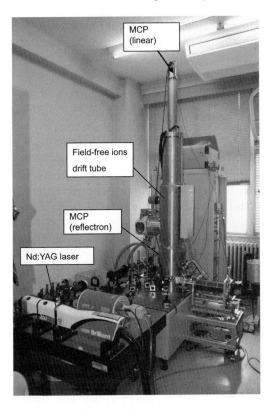

Fig. 3.12. Picture of the time-of-flight mass spectrometer, laser and the optical arrangement employed in the experiment

Fig. 3.13. Modified target plate for the attachment of gold nanorod substrate

The output of the laser was originally at TE polarization and a calcite polarizer was added to increase the polarization extinction ratio. The laser polarization on the target was adjusted using a half-wave plate. The laser spot size on the target substrate was about $\sim$200 μm in diameter. Unless otherwise stated, the mass spectra were acquired at optimized laser fluence which was estimated to be in the range of few ten mJ/cm^2 to $\sim$100 mJ/cm^2. For comparison, the frequency-tripled Nd:YAG laser (355-nm wavelength, not further polarized) was also used to investigate the wavelength dependence. Throughout the experiment, the substrate was scanned, and 40–60 laser shots were used to acquire the mass spectra.

3.4.2 Sample Preparation

All chemicals and analytes were obtained commercially and used without further purification. Bovine insulin was prepared in the aqueous solution of 1% trifluoroacetic acid (TFA). Lys-Lys, Lys-Lys- Lys-Lys-Lys, bradykinin, and melittin were dissolved in water. Lactose was prepared in the aqueous solution of sodium chloride ($\sim$10 ppm) to promote cationization. The citric buffer was prepared by mixing the aqueous solution of citric acid 10 mM with the diammonium citrate (10 mM) at the ratio of 1/2. Working stocks containing the analyte were prepared in the concentration of 1–10 pmol/μl. About 0.2–1 μl of the working stock was pipetted onto the gold nanorod substrate and the droplet was gently dried using a warm air blower. When the droplet was dried, the gold nanorod substrate loaded with the analytes was transferred into the vacuum chamber of the time-of flight mass spectrometer.

3.5 Mass Spectra from Gold Nanostructure

3.5.1 Mass Spectra from Gold-Coated Porous Silicon

Figure 3.14(a) shows the mass spectrum of 5 pmol bradykinin (1060 Da) obtained by irradiating the 532-nm visible laser on the analytes deposited on the gold-coated porous silicon. Besides the protonated ions, $[M + H]^+$, the alkali metal ion adducts, $[M + Na]^+$ and $[M + K]^+$ are also observed in the mass spectrum.

On the bare porous silicon (no gold coating), no molecular ion signal was observed at the same or higher laser fluence (Fig. 3.14(b)). This clearly shows that the gold nanostructure rather than the porous silicon template contributes to the desorption/ionization of the analytes. In an attempt to desorb/ionize the bradykinin from the flat gold surface (~50-nm Au film coated on the flat silicon surface) using the 532-nm laser, only a very weak signal was obtained at high laser power (Fig. 3.14(c)). This indicates that the gold nanostructure was essential in assisting the desorption/ionization of analytes.

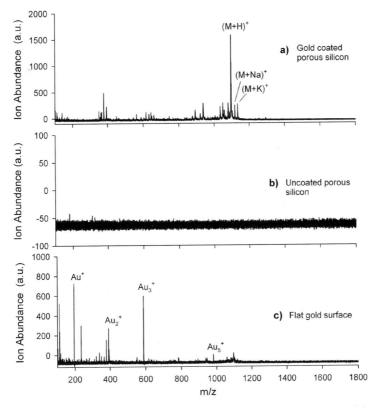

Fig. 3.14. Mass spectrum of bradykinin (1060 Da) obtained by the irradiation of the visible 532-nm laser on **a** gold-coated porous silicon. **b** Uncoated porous silicon. **c** Flat gold surface (~50-nm gold film coated on the flat silicon surface) at higher laser intensity

3.5.2 Mass Spectra from Gold Nanorods

Polarization Dependence

Due to the random structure of the porous silicon substrate, the optical polarization of the incidence light was not well defined. The optical polarization effect was studied using the gold nanorods. Figure 3.15 shows the mass spectra of angiotensin I (1296 Da) obtained using the gold nanorod substrate with different laser polarization.

The laser polarization incidence on the substrate was adjusted using a half-wave plate, without significant change in the incidence laser fluence. The same laser fluence was applied to obtain mass spectra of different laser polarization.

For the shorter nanorods with length <50 nm, the molecular ion signal became maximum when the laser was TM-polarized, and minimum when TE-polarized (Figs. 3.15(a) and 3.15(b)). These short nanorods behave optically as particles on the flat surface, and the dependence of optical absorption on the laser polarization is similar to that of thin gold film. Such optical property could also be influenced by the dielectric alumina template.

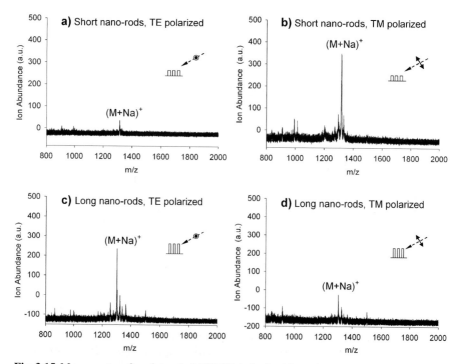

Fig. 3.15. Mass spectra of angiotensin I (1296 Da) obtained from the gold nanorods excited by the laser of different optical polarization. **a** and **b**: The optimum optical polarization for short nanorods (length ∼50 nm) was TM. **c** and **d**: The optimum polarization was reversed to TE for longer nanorods (length ∼100 nm)

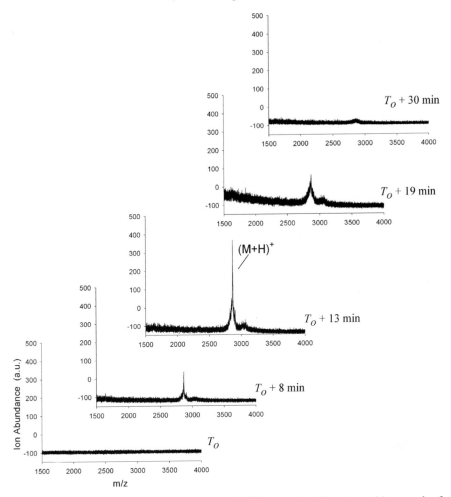

Fig. 3.16. The mass spectra of 5 pmol melittin (2847 Da) acquired from the gold nanorods of different etching time. T_O denotes to the time at which the chemical etching of the alumina template had just reached the gold nanorods, and no ion was detected before T_O. The nanorods started to protrude from the template after T_O and, for this substrate, the ion signal reached maximum at $T_O + 13$ min. Nanorods started to topple after $T_O + 19$ min and the ion signal decreased. After $T_O + 30$ min, some nanorods were detached from the substrate and the melittin ions were difficult to detect

As the length of the gold nanorods increases, (e.g. $\sim$100 nm), the optimum polarization at the excitation wavelength of 532 nm was reversed to TE (Figs. 3.15(c) and 3.15(d)). The reverse of the optimum polarization was due to the excitation of transverse surface plasmon resonance of the gold nanorods. In sum, the obtained mass spectra which are sensitive to the laser polarization agree reasonably with their

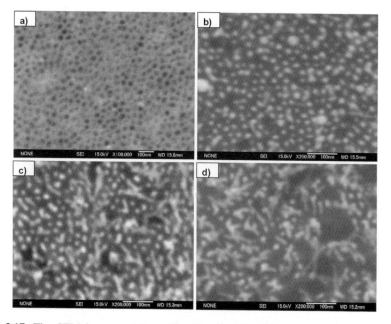

Fig. 3.17. The SEM images of the gold nanorods with different etching. **a** Before T_o, **b** $T_o + 13$ min, **c** $T_o + 19$ min and **d** $T_o + 30$ min

optical properties. In the following experiments, the gold nanorods were ~100 nm in length, and the optical polarization was optimized at TE. As for the gold-coated porous silicon, significant polarization dependence was not observed, but the polarization was arbitrarily maintained at TM.

Post-etching of Gold Nanorods Substrate

The chemical etching of the alumina template was also a key process in obtaining good mass spectrum of the deposited analyte. The desorption/ionization efficiency was found to be quite dependent on the etching time. Figure 3.16 shows the mass spectra of 5 pmol melittin (2847 Da) acquired from the gold nanorods prepared with different etching times. In Fig. 3.16, T_o denotes the time at which the chemical etching of the alumina template just reached the gold nanorods. The SEM images of the gold nanorod substrate taken after the LDI experiment are shown in Fig. 3.17. The analyte ion signals increased as the gold nanorods started to emerge from the template. The optimum condition was achieved at about $T_o + 13$ min when the ~100-nm nanorods were almost completely released from the template. Further etching of the alumina template caused the nanorods to topple (see Fig. 3.17(c)) and the ion signal started to diminish. Excessive etching detached some of the nanorods from the alumina template (Fig. 3.17(d)) and it became difficult to observe ion signals.

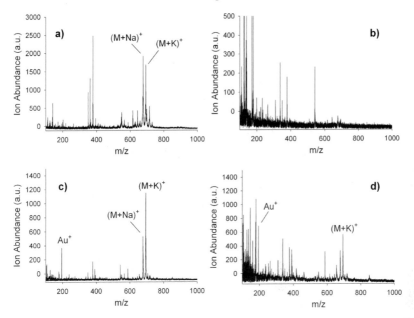

Fig. 3.18. Mass spectra of pentalysine obtained from gold-coated porous silicon using **a** a 532-nm and **b** a 355-nm laser. Mass spectra from gold nanorods using **c** a 532-nm, and **d** a 355-nm laser

Wavelength Dependence

The frequency-doubled (532 nm) and tripled (355 nm) output from the Nd:YAG laser was used to study the laser wavelength dependence. The mass spectra of 5 pmol pentalysine (Lys-Lys-Lys-Lys-Lys, 659 Da) acquired from the gold-coated porous silicon and gold nanorods are shown in Fig. 3.18. The visible LDI using 532-nm laser are depicted in Fig. 3.18(a) and 3.18(c), and the UV-LDI are shown in Fig. 3.18(b) and 3.18(d). The UV laser fluence irradiated on the substrate was estimated to be slightly higher than the visible laser, but accurate comparison was difficult. Instead, the background ions, e.g. Au^+ ions (Fig. 3.18(c) and 3.18(d)), were used as the indirect references for comparison. For both substrates, the 532-nm laser gave better ion yields with less background noise (Fig. 3.18(a) and 3.18(c)). Although ions were also observed with the 355-nm laser, the desorption/ionization was not as efficient and the background ions appeared to be stronger, probably due to the higher laser threshold fluence. The background ions might be due to the contaminants absorbed from the air or vacuum chamber.

Desorption/Ionization of Low Molecular Weight Analytes

Because ions could be produced without the addition of a chemical matrix, the present method may find application in the analysis of low molecular weight analytes, which are rather difficult to handle in the conventional MALDI. Figures 3.19(a),

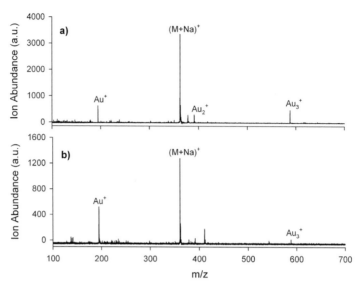

Fig. 3.19. Desorption/ionization of lactose (342 Da) **a** from the gold nanorods using a 532-nm laser, and **b** from the gold-coated porous silicon using a 532-nm laser

and 3.19(b) show, respectively, the mass spectra of 5 pmol lactose (342 Da) acquired from the gold nanorods and gold coated porous silicon by using a 532-nm laser. Both substrates produced similar mass spectra. Although gold atoms were consumed during the laser desorption/ionization, no obvious damage was observed on the substrate at the moderate laser fluence, which was the threshold for most of the peptides and small carbohydrates. The gold cluster ions, Au_n^+, were easily identified and were used as the references for mass calibration.

The mass spectra for ~6 pmol Lys-Lys (274 Da) acquired from the gold nanorods and stainless steel target plate by using a 532-nm laser are shown in Fig. 3.20. At high laser fluence, Lys-Lys ion signals were also observable from the stainless steel (Fig. 3.20(b)), but the most intense and homogeneous ion signals were produced from the gold nanorods (Fig. 3.20(a)). Besides Au^+, some background ions (<300 m/z) were also present in the low mass region, which were probably due to contaminants and hydrocarbons which were absorbed by the nanorods from the air or vacuum.

Addition of Proton Source

Probably due to the lack of acid or proton source on the gold nanostructured substrate, the protonation of peptide and protein was not as efficient as the MALDI method. Besides glycerol (which is the typically used liquid matrix), it has been recently reported that the citric buffer (mixture of citric acid and ammonium citrate) could also work efficiently as the proton donor [16].

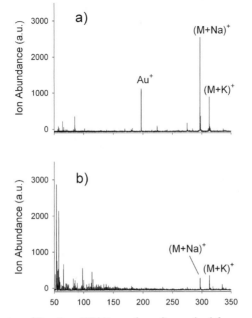

Fig. 3.20. Mass spectra of Lys-Lys (274 Da, ~6 pmol) acquired from **a** gold nanorods, and **b** stainless steel using a 532-nm laser

The visible laser desorption/ionization of 650 fmol bradykinin and 2 pmol bovine insulin from the gold nanorods with the addition of the citric buffer during sample preparation are shown in Figs. 3.21(a) and 3.21(b), respectively. The molar ratio of analyte to citric acid was ~1/350. The citric buffer significantly improved the analytes' ion yields, and the resolution of the mass spectra. The ions contributed by the citric buffer were also observed in the mass spectra. The peaks at m/z 215 and 231 (in Fig. 3.21(a)) correspond to the sodium and potassium ion adducts of the citric acid ($[M + Na]^+$, and $[M + K]^+$). The yet-determined background peak at m/z 183 was probably due to the contaminant absorbed on the gold nanorods.

For the two-phase matrix, the nanopowders have to be suspended in liquid glycerol to prevent agglomeration, and the ion signals deteriorate after complete evaporation of glycerol in the vacuum [12]. In contrast, the nanostructures allow the use of various solid proton sources. Besides citric acid, several compounds with suitable hydroxyl groups, e.g. tartaric acid, xylitol and mannitol were also found to be useful in analyte ionization.

Surface-Enhanced Raman Spectroscopy (SERS) from Gold Nanorods

In a separate experiment, we demonstrated that the gold nanorods used in our experiments were also SERS-active. The Raman experiment was conducted using a Jasco NRS-2100 micro-Raman monochromator spectrometer. The analyte, rhodamine 6G

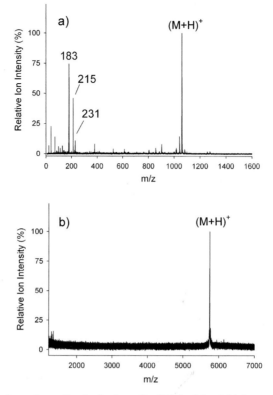

Fig. 3.21. Visible laser desorption/ionization of **a** 650 fmol bradykinin, and **b** 2 pmol bovine insulin from the gold nanorods with addition of citric buffer

was dissolved in water and a continuous-wave 488-nm argon laser was used for excitation. The operating laser power was about 0.35 mW and the exposure time was 10 s. The Raman spectroscopy of 5 pmol rhodamine 6G (R6G) deposited on the gold nanorods is shown in Fig. 3.22(a). For comparison, the signal obtained from the aluminum substrate under the same experimental condition is shown in Fig. 3.22(b) and the respective mass spectra acquired by using 532-nm laser are shown in Fig. 3.22(c) and 3.22(d).

3.5.3 Gold Nanoparticle-Assisted Excitation of UV-absorbing MALDI Matrix by Visible Laser

In UV-MALDI, the photo-ionization routes to produce radical ions include direct two-photon ionization [8], and exciton pooling of the excited matrix molecules [9]. The ionization potential (IP) for DHB is 8 eV and the energy for first excited state is 3.466 eV [64]. Because the IP is still higher for the two-photon energy from the typical UV laser (e.g., nitrogen laser), the photo/thermal hypothesis was proposed [65], where the energy deficit is made up by the thermal energy. For the exciton pooling,

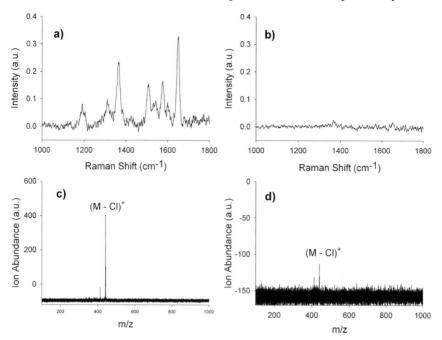

Fig. 3.22. Raman signals of 5 pmol Rhodamine 6G (471 Da) obtained from **a** the gold nanorods, and **b** aluminum substrate. The excitation wavelength was 488 nm. The prominent peaks are at 1365, 1510, 1575 and 16550 cm^{-1}. **c** and **d** are the respective mass spectra acquired by using a 532-nm laser. The R6G ions appeared as $(M-Cl)^+$

it is readily accessible by one UV photon, and two or more excited matrix molecules pool their energies to form radical ions or higher excited states of which the transfer of proton to the target analytes takes place.

In this section, we demonstrate the excitation of UV absorbing MALDI matrix and the desorption/ionization of the doped biomolecular analytes by using a frequency-doubled Nd:YAG laser (532 nm), assisted by gold nanoparticles. In our experiment, Au thin film ($\sim$10 nm) was first coated on the matrix surface and the Au nanoparticles were prepared and deposited directly on the matrix surface by ablating the Au film in the vacuum. Coating the biosample with Au thin film is common in the secondary ion mass spectrometry (SIMS) [60] and, recently, metallic coating has also been reported to improve the UV-MALDI imaging [61]. However, in our experiment, the bulk Au film on the laser spot will be ablated, and the remaining Au nanoparticles were used to excite the matrix molecule using visible laser, of which the photon energy is insufficient for the free molecules.

Laser Ablation of the Au Film

The experiment began by ablating the Au film coated on the MALDI matrix to form nanoparticles. The laser fluence was adjusted to the ablation threshold and the Au

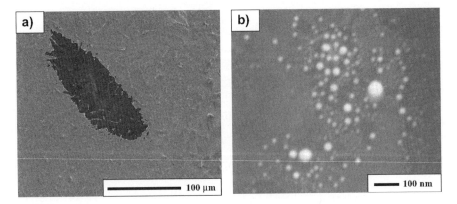

Fig. 3.23. SEM images showing the ablation of Au by the Nd:YAG laser. **a** The ablation of the bulk Au thin film. **b** A close-up on the ablated region showing the formation of Au nanoparticles by laser ablation

cluster ions were monitored directly from the mass spectrum. As the Au thin film was of ∼10 nm, the laser threshold fluence for ablation was much lower compared to that of bulk Au. The laser spot size on the target was about ∼0.1 mm in diameter and the operating laser fluence was estimated to be 50–100 mJ/cm^2. After few laser shots, the Au ions were significantly reduced, and the mass spectra were acquired directly or with slight adjustment of the laser fluence. Unless otherwise stated, the mass spectra for the matrix and the analytes were usually acquired after 5–10 initial laser shots, and from the accumulation of 30–100 single-shot mass spectra on the same laser spot.

The scanning electron micrograph in Fig. 3.23(a) shows the ablation of Au thin film by the Nd:YAG laser after several laser shots. A close-up inspection on the ablated region showed gold nanoparticles with size ranging from <10 nm to ∼100 nm (see Fig. 3.23(b)).

Several processes could contribute to the formation of the gold nanoparticles, e.g., the melting of the Au film to form thermodynamically stable particles, or the redeposition of the ablated Au clusters back onto the surface. Some nanoparticles could also be embedded into the matrix after ablation.

Radical Matrix Ions Produced by Visible Laser

The typical first laser shot mass spectrum is shown in Fig. 3.24(a). The first laser shot ablated most of the Au atoms and the mass spectrum was dominated by intense Au cluster ions. Because the Au film was deposited on the matrix surface, the laser ablation Au film was less likely to assist the desorption of the underneath matrix. Only very weak or no matrix ion signal was observed after the first few laser shots. Figure 3.24(b) shows the Au nanoparticles assisted visible-MALDI mass spectrum of the super-DHB (2,5-Dihydroxybenzoic acid [DHB, 154 Da] and 2-hydroxy-5-methoxybenzoic acid [HMB, 168 Da]).

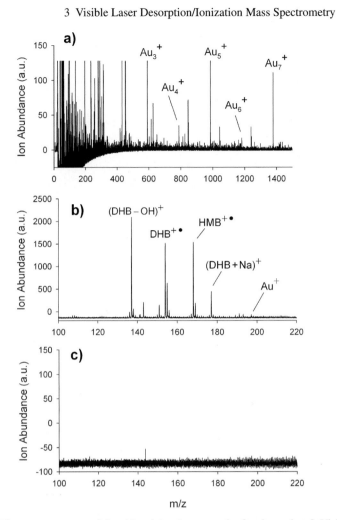

Fig. 3.24. a The mass spectrum of the ablated Au clusters at the first laser shot. **b** Visible laser desorption/ionization of super-DHB assisted by gold nanoparticles. **c** No observable matrix signal was obtained without gold nanoparticles using the visible laser

The desorbed ions consist of $DHB^{+\bullet}$, $[DHB + H]^+$, $[DHB + Na]^+$, $[DHB - OH]^+$, $HMB^{+\bullet}$, and $[HMB + H]^+$, which are the typical matrix related ions observed in UV-MALDI. Gold ions Au^+ also appeared in the mass spectrum, but in relatively lower abundance. Irradiating the matrix without Au coating produced no observable matrix ions (see Fig. 3.24(c)).

The analyte and the matrix related ions became apparent after the fifth laser shot as the Au clusters ions reduced. Figure 3.25 shows the mass spectrum of the bradykinin (1060 Da) acquired by accumulation of 26 single-shot spectra from the same laser spot. The matrix and analyte ions usually lasted for more than 50 laser shots.

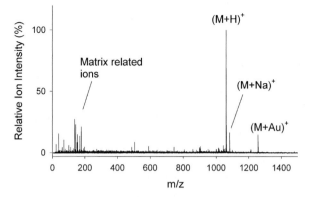

Fig. 3.25. Au nanoparticles assisted MALDI mass spectrum of bradykinin (accumulation of 26 laser shots)

3.6 Discussion and Conclusion

For the silicon-based substrate, the macro-porous structure, or grooves of ~100 nm (but not smaller) are required to produce a good ion signal [17, 19]. Since the nanorods, which are ~15 nm in diameter, work equally well with the gold-coated porous silicon, the macro-porous feature is not essentially required in the case of gold. Due to the inert nature of gold, the nanorod substrate can be kept in the atmospheric environment and reused after proper cleaning, without significant deterioration in performance.

Regarding the desorption/ionization mechanism, because the gold porous silicon and nanorods essentially resemble the finely divided particles, it may function as the two-phase matrix (inorganic particles in liquid suspension), in which the rapid heating of the substrate and the resultant peak surface temperature leads to the desorption/ionization of the residual solvent molecules that may subsequently assist in the formation of analyte ions [62, 63]. The observed alkali ion adducts in some of the spectra also suggest that the ions were formed by the gas phase cationization which are likely thermal-driven.

The sensitivity of the LDI was also found to be enhanced with the addition of a proton source. Besides citric acid and glycerol (which have been frequently used in MALDI), we found that certain sugar alcohols (e.g., xylitol and mannitol), and tartaric acid are also useful as proton donors. In the case where a proton source was added, they were excited by the nanostructure as their optical absorption is almost negligible in the near-UV and visible region, and the ionization of analytes probably took place in the gas phase. The excitation mechanism remains unknown, and it is usually assumed that the heating is the only process induced by the nanoparticles or nanostructure upon photon absorption. However, photo-heating and the peak temperature might not be the exclusive mechanism for ion desorption and localized effects, such as an enhanced electric field, have also been proposed [63]. As seen from the laser wavelength and polarization dependence, the optimum condition of the desorption/ionization was achieved when the surface plasmon resonance is excited. A certain degree of optical-field enhancement is also expected at the plasmon resonance as we had obtained the surface-enhanced Raman spectrum of rhodamine 6G from the gold nanorods. However, it remains uncertain whether this is merely due to the plasmon-enhanced optical absorption.

In the visible MALDI experiment, we showed that with the presence of Au nanoparticles, the UV-MALDI matrix could also be photo-ionized by the 532-nm laser even though the photon energy is insufficient for free molecules.

Since the 532-nm laser (2.33 eV) was employed in our experiment, at least three or more photons are needed for direct photo/thermal multiphoton process, and two photons for excitation. Although the presence of Au nanoparticles increase the thermal heating of the matrix molecules, which could lead to the phase evaporation and molecular desorption, it remains questionable if the thermal process is sufficient for the excitation and ionization of the matrix molecules. Thus, in addition to heating effect, a combination of several processes could contribute to the observed ions. As gold nanoparticles couple the photon to the surface plasmon efficiently, the induced electromagnetic field enhancement—which promotes nonlinear optical processes—introduces the possibility of two or multiple photon excitation of the matrix molecules. The chemical enhancement in the SERS, of which the optical absorption band was broadened or red-shifted due to the interaction between absorbates and conduction electrons on the metal surface, also allows the excitation of matrix molecules by the visible laser.

We note that the optical near-field enhancement is active only in the vicinity of the nanostructure and is characterized by the penetration depth corresponding to the size of the nanoparticles [66]. Thus, a metallic tip sharpened to subwavelength radius may be used as a scanning nanoprobe to desorb the biomolecules with nanometer resolution. The Au nanoparticles may also find application in selective ionization, of which only the specific sides of the biosample that are bound to the particles are ionized by the laser.

References

[1] F.H. Field, J.L. Franklin, *Electron Impact Phenomena and the Properties of Gaseous Ions*, 1st edn. (Academic Press, New York, 1957)
[2] J.B. Fenn, M. Mann, C.K. Meng, S.F. Wong, C.M. Whitehouse, Science **246**, 64 (1989)
[3] K. Tanaka, H. Waki, Y. Ido, S. Akita, Y. Yoshida, T. Yoshida, T. Matsuo, Rapid Commun. Mass Spectrom. **2**, 151 (1988)
[4] M. Karas, F. Hillenkamp, Anal. Chem. **60**, 2299 (1988)
[5] F. Hillenkamp, J. Peter-Katalinić, *MALDI MS* (Wiley-VCH, Weinheim, 2007)
[6] J.H. Gross, *Mass Spectrometry* (Springer, Berlin, 2004)
[7] T.J.P. Naven, D.J. Harvey, Rapid Commun. Mass Spectrom. **10**, 829 (1996)
[8] H. Ehring, M. Karas, F. Hillenkamp, Org. Mass Spectrom. **27**, 472 (1992)
[9] R. Zenobi, R. Knochenmuss, Mass Spectrom. Rev. **17**, 337 (1998)
[10] D.S. Cornett, M.A. Duncan, I.J. Amster, Anal. Chem. **65**, 2608 (1993)
[11] C.J. Smith, S.Y. Chang, E.S. Yeung, J. Mass Spectrom. **30**, 1765 (1995)

[12] M. Schürenberg, K. Dreisewerd, F. Hillenkamp, Anal. Chem. **71**, 221 (1999)

[13] T. Kinumi, T. Saisu, M. Takayama, H. Niwa, J. Mass Spectrom. **35**, 417 (2000)

[14] J.A. McLean, K.A. Stumpo, D.H. Russell, J. Am. Chem. Soc. **127**, 5304 (2005)

[15] J. Sunner, E. Dratz, Y.C. Chen, Anal. Chem. **67**, 4335 (1995)

[16] C.-T. Chen, Y.-C. Chen, Rapid Commun. Mass Spectrom. **18**, 1956 (2004)

[17] S. Okuno, R. Arakawa, K. Okamoto, Y. Matsui, S. Seki, T. Kozawa, S. Tagawa, Y. Wada, Anal. Chem. **77**, 5364 (2005)

[18] J. Wei, J.M. Buriak, G. Siuzdak, Nature **399**, 243 (1999)

[19] Z. Shen, J.J. Thomas, C. Averbuj, K.M. Broo, M. Engelhard, J.E. Crowell, M.G. Finn, G. Siuzdak, Anal. Chem. **73**, 612 (2001)

[20] E.P. Go, J.V. Apon, G. Luo, A. Saghatelian, R.H. Daniels, V. Sahi, R. Dubrow, B.F. Cravatt, A. Vertes, G. Siuzdak, Anal. Chem. **77**, 1641 (2005)

[21] J.D. Cuiffi, D.J. Hayes, S.J. Fonash, K.N. Brown, A.D. Jones, Anal. Chem. **73**, 1292 (2001)

[22] N.H. Finkel, B.G. Prevo, O.D. Velev, L. He, Anal. Chem. **77**, 1088 (2005)

[23] S.H. Bhattacharya, T.J. Raiford, K.K. Murray, Anal. Chem. **74**, 2228 (2002)

[24] R.H. Ritchie, Surf. Sci. **34**, 1 (1973)

[25] H.C. van de Hulst, *Light Scattering by Small Particles* (Dover, New York, 1981)

[26] H. Du, Appl. Opt. **43**, 1951 (2004)

[27] U. Kreibig, M. Vollmer, *Optical Properties of Metal Clusters* (Springer, Berlin, 1995)

[28] S. Link, M.A. El-Sayed, Int. Rev. Phys. Chem. **19**, 409 (2000)

[29] M. Moskovits, Rev. Mod. Phys. **57**, 783 (1985)

[30] R. Antoine, M. Pellarin, B. Palpant, M. Broyer, B. Prével, P. Galletto, P.F. Brevet, H.H. Girault, J. Appl. Phys. **84**, 4532 (1998)

[31] M. Danckwerts, L. Novotny, Phys. Rev. Lett. **98**, 026104 (2007)

[32] K. Arya, R. Zeyher, Phys. Rev. B **24**, 1852 (1981)

[33] A. Campion, J.E. Ivanecky III, C.M. Child, M. Foster, J. Am. Chem. Soc. **117**, 11807 (1995)

[34] P.G. Kik, S.A. Maier, H.A. Atwater, Mat. Res. Soc. Symp. Proc. **705**, Y3.6 (2001)

[35] S.A. Maier, P.G. Kik, H.A. Atwater, S. Meltzer, E. Harel, B.E. Koel, A.A.G. Requicha, Nature Mater. **2**, 229 (2003)

[36] D.J. Maxwell, J.R. Taylor, S. Nie, J. Am. Chem. Soc. **124**, 9606 (2002)

[37] Y.C. Cao, R. Jin, C.A. Mirkin, Science **297**, 1536 (2002)

[38] I. El-Sayed, X. Huang, M. El-Sayed, Cancer Lett. **239**, 129 (2006)

[39] M.J. Shea, R.N. Compton, Phys. Rev. B **47**, 9967 (1993)

[40] I. Lee, T.A. Callcott, E.T. Arakawa, Phys. Rev. B **47**, 6661 (1993)

[41] W. Hoheisel, K. Jungmann, M. Vollmer, R. Weidenauer, F. Trger, Phys. Rev. Lett. **60**, 1649 (1988)

[42] J. Brewer, H.G. Rubahn, Chem. Phys. **303**, 1 (2004)

[43] R.H. Ritchie, J.R. Manson, P.M. Echenique, Phys. Rev. B **49**, 2963 (1994)

[44] E. Hutter, J.H. Fendler, Adv. Mater. **16**, 1685 (2004)

[45] P. Englebienne, A.V. Hoonacker, M. Verhas, Spectroscopy **17**, 255 (2003)

[46] L.C. Chen, T. Ueda, M. Sagisaka, H. Hori, K. Hiraoka, J. Phys. Chem. C **111**, 2409 (2007)

[47] L.C. Chen, J. Yonehama, T. Ueda, H. Hori, K. Hiraoka, J. Mass Spectrom. **42**, 346 (2007)

[48] L.T. Canham, Appl. Phys. Lett. **57**, 1046 (1990)

[49] O. Bisi, S. Ossicini, L. Pavesi, Surf. Sci. Rep. **38**, 1 (2000)

[50] M.S. Sander, L.-S. Tan, Adv. Funct. Mater. **13**, 393 (2003)

[51] J. Pérez-Juste, I. Pastoriza-Santos, L.M. Liz-Marzán, P. Mulvaney, Coord. Chem. Rev. **249**, 1870 (2005)

[52] H. Masuda, K. Yada, A. Osaka, Jpn. J. Appl. Phys. **37**, L1340 (1998)
[53] O. Jessensky, F. Müller, U. Gösele, Appl. Phys. Lett. **72**, 1173 (1998)
[54] H. Masuda, K. Fukuda, Science **268**, 1466 (1995)
[55] H. Masuda, M. Satoh, Jpn. J. Appl. Phys. **35**, L126 (1996)
[56] S. Ono, N. Masuko, Surf. Coat. Technol. **169**, 139 (2003)
[57] C.A. Foss, G.L. Hornyak, Jr., J.A. Stockert, C.R. Martin, J. Phys. Chem. **96**, 7497 (1992)
[58] C.R. Martin, Science **266**, 1961 (1994)
[59] M. Born, E. Wolf, *Principles of Optics*, 2nd edn. (Pergamon, Oxford, 1964)
[60] A. Delcorte, N. Medard, P. Bertrand, Anal. Chem. **74**, 4955 (2002)
[61] G. McCombie, R. Knochenmuss, J. Am. Soc. Mass Spectrom. **17**, 737 (2006)
[62] S. Zumbuhl, R. Knochenmuss, S. Wulfert, F. Dubois, M.J. Dale, R. Zenobi, Anal. Chem. **70**, 707 (1998)
[63] S. Alimpiev, S. Nikoforov, V. Karavanskii, T. Minton, J. Sunner, J. Chem. Phys. **115**, 1891 (2001)
[64] V. Karbach, R. Knochenmuss, Rapid Commun. Mass Spectrom. **12**, 968 (1998)
[65] D.A. Allwood, P.E. Dyer, R.W. Dreyfus, Rapid Commun. Mass Spectrom. **11**, 499 (1997)
[66] M. Ohtsu, H. Hori, *Near-Field Nano-Optics* (Kluwer Academic/Plenum, New York, 1999)

4

Near-Field Optical Photolithography

M. Naya

4.1 Introduction

The evolution of photolithography technology has been supported particularly by the advance of reduced projection exposure technology and resist technology. The performance of reduced projection exposure technology mainly depends upon two basic parameters, i.e., the resolution, RP, and the depth of focus, DOP. If the exposure wavelength for the projection optical system is λ, and the numerical aperture of the projection lens is NA, the two basic parameters are expressed by

$$RP = \frac{k\lambda}{n\mathrm{NA}},$$ (4.3)

and

$$DOP = \frac{kn\lambda}{2\mathrm{NA}^2},$$ (4.4)

where n is the refraction index of the projected medium. In order to improve the resolution for lithography, it is essential to reduce the wavelength λ and to increase the numerical aperture of the projection lens, NA. If NA is increased, the resolution is improved, but the depth of focus is reduced in inverse proportion to the square of NA.

To improve the resolution, the exposure wavelength λ has been shortened from that for the g-line (436 nm) to that for the i-line (365 nm), and at present, the excimer laser (248 nm, 193 nm) has become the most popular. However, with lithography using light, the diffraction limit for light provides the limit of resolution, and it is generally accepted that, if an F_2 excimer laser with a wavelength of 248 nm is used, a fine pattern of 100 nm in line width is the limit of lithography using a lens series optical system. If an attempt is made to provide a resolution on the order of less than 100 nm, electron beam or X-ray (particularly SOR light, i.e., synchrotron orbital radiation) lithography technology must be used. Electron beam lithography can control the formation of a pattern on the order of nanometers with high accuracy, providing a significantly greater depth of focus than that obtained with the optical system. In addition, it offers an advantage in that it can directly draw a figure on the wafer without a mask. There is a drawback, however. Because the throughput is low, and the cost is high, electron beam lithography is not suited to volume production. X-ray lithography can provide an approximately one digit higher resolution and accuracy than can be obtained with the excimer laser lithography, either when full-scale exposure is carried out with a 1-to-1 mask or when a reflection-type image formation optical system is used for exposure. However, X-ray lithography presents problems in that the mask is difficult to prepare, the feasibility is low, and the cost is high due to the device. With lithography using an electron beam or X-ray, a resist must be developed in accordance with the exposure method, and problems still exist with respect to sensitivity, resolution, resistance to etching, etc. Furthermore, these technologies include several other problems:

(1) When the wavelength of light becomes short, special glass materials are required for conventional optical systems, including the lens, since conventional glasses are not adaptable.
(2) It is necessary to develop new resist materials that respond to short wavelength.
(3) Developing a new system is both expensive and time-consuming.

Since the near-field optical lithography (NFOL) using near-field light is not affected by the diffraction limit of light, it holds promise for realizing high-resolution lithography using conventional inexpensive light sources, optical systems and resist materials [1–15]. The scanning near-field optical probe method [1–3] and the cell projection methods which use masks with apertures [4–15], have been reported as near-field optical photolithography methods. From the viewpoint of throughput, the cell projection method is considered to be more suitable for practical application.

4.2 Near-Field Optical Photolithography (NFOL)

4.2.1 Principle of NFOL

Figure 4.1 shows the principle of the cell projection type NFOL. The master mask pattern is formed onto the metal thin film on the optical transparent substrate. The mask is in hard contact with the resist layer on the substrate. Exposing light generates the near-field light around the aperture of mask. The near-field light emitted from the mask exposes the resist layer, and the mask pattern is transferred to it through the developing process. Subsequently, the mask pattern is transferred to the resist.

4.2.2 NFOL with Bilayer Resist Process

Because the penetration depth of near-field light is small, NFOL faces the difficulty of achieving a sufficient etching depth, i.e., a high aspect ratio. In order to obtain high aspect ratio patterning, we developed a bilayer resist process [12], which is

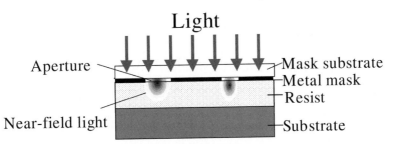

Fig. 4.1. Principle of near-field optical lithography. The mask is in hard contact with the resist layer on the substrate. Exposing light generates the near-field light around the aperture of mask. The near-field light emitted from the mask exposes the resist layer, and the mask pattern is transferred to it through the developing process. Subsequently, the mask pattern is transferred to the resist

one of the surface layer imaging methods [16–18]. Figure 4.2 shows the principle of near-field optical photolithography using a bilayer resist. First, a nonphotosensitive and plasma-etchable resist material layer (the bottom resist layer) is formed on the surface of the substrate. The thickness of the bottom resist layer is such that it can sufficiently withstand the working process. A photosensitive dry etching-resistant material layer (the upper resist layer) is formed on the bottom resist layer. The thickness of the upper resist layer is sufficiently thinner than the penetration depth of near-field light. At the exit plane of the mask, the range of the near-field light shrinks as the width of the slit is reduced. Therefore, for smaller width, thinner resist is required. The mask makes hard contact with the resist layer on the substrate. The light of a wavelength, to which the upper resist layer is photosensitive, is illuminated above the mask. The upper resist layer is exposed to the near-field light emitted from the mask, and the mask pattern is transferred to it through the developing process. This upper resist pattern is transferred to the bottom resist layer through the dry-etching process in which the pattern of the upper resist layer serves as a mask to the dry-etching process. This process achieves high aspect ratio in NFOL.

4.3 Experiments

4.3.1 Experimental Set-up

Figure 4.3 shows our hand-made, hard-contact exposure experimental device used in this experiment. The light source is a mercury lamp. The UV filter cut i-line (373 nm) and other shorter wavelength light, and only g-line (436 nm) was exposed. The polarizer was used to control the polarization of exposing light. To obtain an extremely hard contact between the mask and the resist surfaces, we devised a vacuum hard-contact system.

As the photo-mask for the near-field illumination, a slit pattern written on a Cr film (40-nm thick) on silica glass by electron beams was used. Figure 4.4 shows an

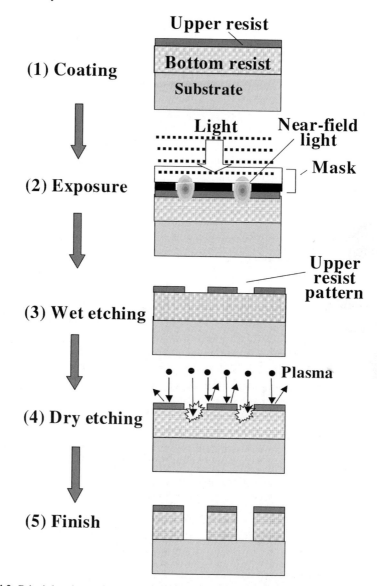

Fig. 4.2. Principle of near-field photolithography employing the two-layer resist. Near-field light is illuminated to form a fine structure not larger than the wavelength of the illuminating light in the upper resist layer. Subsequently, a pattern that is formed in the upper resist layer is transferred to the bottom layer by the plasma dry-etching process. Hence, this near-field photolithography can achieve a high aspect ratio that has a high definition not larger than the wavelength and beyond the range of the near-field light

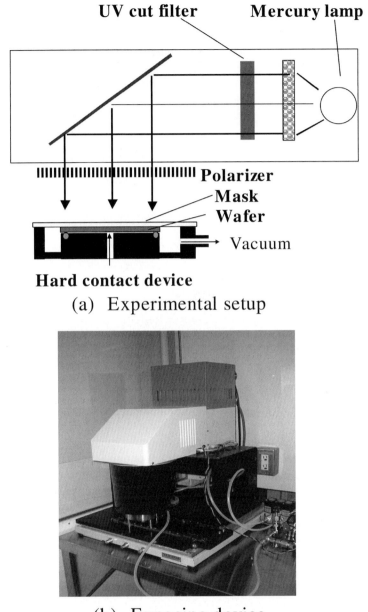

UV cut filter   **Mercury lamp**

Polarizer
Mask
Wafer
Vacuum

Hard contact device
(a) Experimental setup

(b) Exposing device

Fig. 4.3. Near-field pattern exposure system. The mask is brought into intimate contact with the resist surface, and a g-line light emitted from a mercury lamp is illuminated

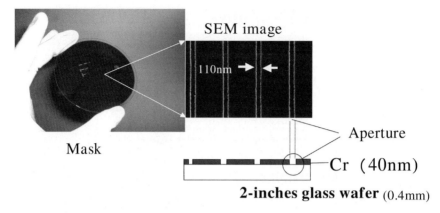

Fig. 4.4. Mask pattern. Chrome slit apertures with a width of 110 nm are formed on the glass wafer

SEM image of the photo-mask. The slit width shown in the figure is 110 nm and the pitch is 1 μm.

4.3.2 Patterning Experiment of Monolayer Resist

This section discusses the experimental result of the monolayer resist. In this experiment, photo-resist (TSMR8900) was spin-coated to a thickness of 1000 nm on the silicon wafer. After spin coating, it was baked for 20 min at 100 °C. Next, the mask was hard contacted on the surface of resist, and light was exposed. For this experiment, the intensity of exposed light was 37 mW/cm^2, and the exposure time was 25 s. After exposure, the wafer was dipped in the developer NMD-3 for 60 s, and washed with water. Through this process, the mask pattern was transferred to the surface of resist. Figure 4.5 is the SEM image of the transferred lattice pattern. This image shows that the width of the slit was 180 nm. It was sufficiently smaller than the wavelength of exposed light. However, the shape of the pattern was spherical and the depth was only 180 nm. Such a pattern is impossible to use as the resist pattern for dry etching the substrate.

4.3.3 Patterning Experiment of Bilayer Resist Process

In the case of bilayer resist materials, FH-SP3CL is used for the upper resist layer and FHi028 for the bottom resist layer (both materials are manufactured at Fujifilm Arch Co., Ltd). The upper resist, FH-SP3CL, is a silicon-containing positive resist developed for the bilayer resist process. This resist is photosensitive in the case of the g- and i-lines and can be easily patterned by the ordinal wet-etching process. However, due to the exposure of O$_2$ plasma, Si contained in this material changes to SiO$_2$ and becomes an etching-resist for O$_2$ plasma. The bottom resist FHi028 is changed to a nonphotosensitive layer by baking, and it is then ready to be etched by O$_2$ plasma. The process of bilayer resist is as follows. First, FHi028 was spin-coated

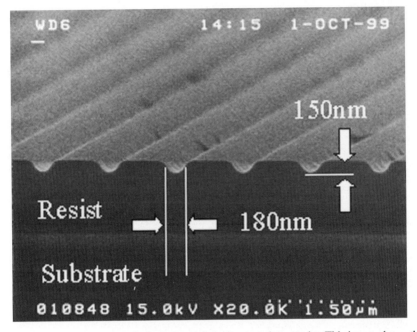

Fig. 4.5. Transferred pattern of a 110-nm mask in the monolayer resist. This image shows that the width of slit was 180 nm. It was sufficiently smaller than the wavelength of exposed light. However, the shape of the pattern was spherical and the depth was only 180 nm. Such a pattern is impossible to use as the resist pattern for dry etching the substrate

over the silicon substrate to form a 480-nm thick bottom resist layer and baked for 90 min at 200 °C to render it nonphotosensitive. Next, PH-SP3CL was spin-coated over the surface of the bottom resist layer and baked for 20 min at 100 °C to form a 70-nm upper resist layer. After forming the resist layer the mask was put into hard contact with the upper resist layer. Subsequently, a g-line light was illuminated. In this experiment, the polarization of exposure light was parallel to the slit (TE), and the exposure time was 250 s. Following the illumination, the bilayer resist structure was dipped into a developer NMD-3 for 60 s in order to transfer the mask pattern on the upper resist layer. After washing with water, it was baked for 60 min at 100 °C. Further, the pattern formed on the upper resist layer was transferred to the bottom resist layer through the mask pattern for the O_2 plasma dry-etching process. The dry-etching system used in this experiment was DES-215R, manufactured by Plasma Systems Co., Ltd. The etching was conducted under the following conditions: 7 mtorr vacuum, 150 W RF power, −10 °C stage temperature, and 254 s etching time. The result of the transferred slit pattern with a line width of 110 nm is shown in Fig. 4.6. We can observe transferred slit patterns having a line width of 130 nm and a depth of 550 nm. This width is smaller than 1/3 of the wavelengths of the exposed light. It is of great significance that the pattern transfer has been achieved using a conventional, inexpensive light source emitting a g-line. This fact

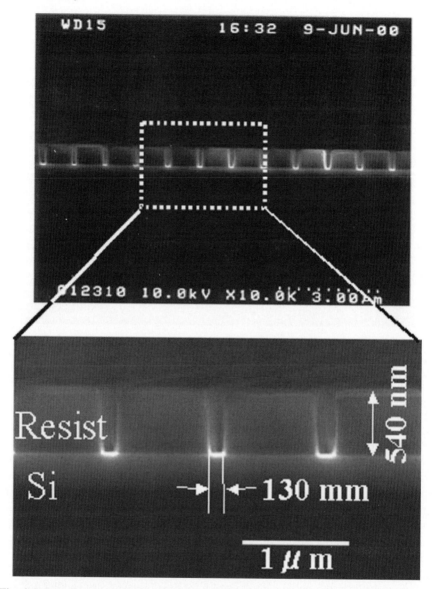

Fig. 4.6. Transferred pattern of a 110-nm mask in the bilayer resist. The result of the transferred slit pattern with a line width of 110 nm is shown. We can observe transferred slit patterns having a line width of 130 nm and a depth of 550 nm. This width is smaller than 1/3 of the wavelengths of the exposed light

suggests that a high resolution and high aspect ratio can be achieved in photolithography through a combination of near-field optical lithography and the bilayer resist technology.

4.4 Simulations

4.4.1 Dependency of Thickness of Resist Layer

Because of near-field optical property, the transferred pattern of the NFOL does not correspond to the mask's pattern. In order to study this, we simulated the electric field in the resist using a three-dimensional finite-difference time-domain (FDTD) method. Simulations were performed for chrome transmission grating of a 1-micron pitch, with a width of 110 nm and a thickness of 40 nm on a glass substrate ($n = 1.52$); the mask contacts with the resist ($n = 1.7$) of thickness t on a Si substrate ($n = 4.831, k = 0.185$). For these calculations, x and y was set as the periodical boundary condition and z was set as the absorption boundary condition. The polarization of the electrical field is parallel to the slit pattern. Figure 4.7(a) shows the distribution of the electrical field when t is infinite. In this case, near-field light penetrates deeper than 70 nm, which is the same as the thickness of the upper resist layer in this experiment. Moreover, we can observe the far-field pattern of light caused by the interference with far-field light radiated from neighboring slits. Figure 4.7(b) shows the distribution of the electrical field with resist thickness of 540 nm. The near-field light penetrates deeper than 70 nm in this configuration as well; however, this pattern spread wider than the slit width. Moreover, there is a standing wave pattern besides the near-field light. These patterns are the light standing wave between the mask and the silicon substrate. It is believed that this pattern decreases in resolution. Figure 4.7(c) shows the distribution of the electrical field in a resist thickness of 70 nm. This figure shows only the near-field pattern localized around the aperture. Furthermore, Fig. 4.7(d), in a resist thickness of 40 nm, shows that the penetration depth of near-field light is shorter than the resist thickness of 70 nm. We think the reason for this phenomenon is as follows. When the polarization of incident light is parallel to the slit pattern, the polarization of near field light is parallel to the slit, and also parallel to the substrate. This field generates a par-

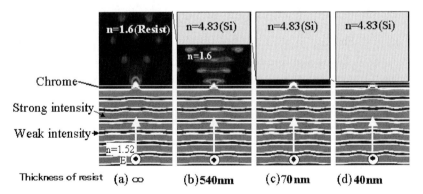

Fig. 4.7. a Distribution of the electrical field when t is infinite, **b** a resist thickness of 540 nm, **c** distribution of the electrical field in a resist thickness of 70 nm and **d** a resist thickness of 40 nm

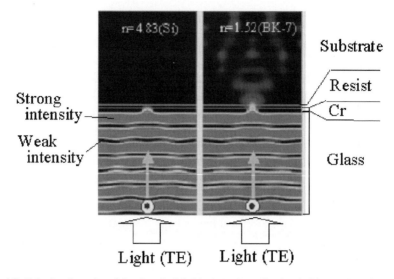

Fig. 4.8. Calculated results of the electrical field where the reflection index number of substrate is 4.83 (*left*) and 1.52 (*right*)

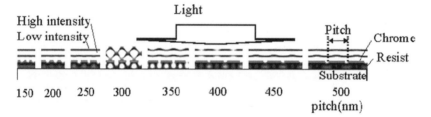

Fig. 4.9. The result of the simulating the electric field in the resist layer generated from the slit of a different pitch. The optical patterns in the resist are different and depend on the pitch of the slits

allel but opposite polarization electrical field in the substrate if the reflection index of substrate is higher than the resist. As a result, both electrical fields erase each other, and that becomes zero on the substrate's surface. To confirm this assumption, we simulated the model where the reflection index number of the substrate is 1.52; it is mostly the same as the resist layer. Figure 4.8 shows the result of this calculation. Here, near-field light reaches the substrate. This result confirms our assumption. These results indicate that a mask must be designed that considers the thickness of the resist.

4.4.2 Dependency of Pitch

Figure 4.9 shows the results of simulating the electric field in the resist layer generated from the slit of a different pitch. These results show that the optical patterns

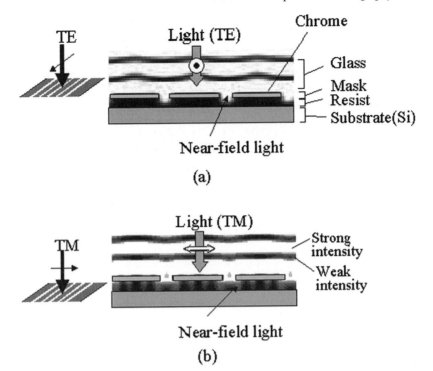

Fig. 4.10. The result of simulating the electric field in the resist layer generated from the slit when the polarization of exposure light was **a** parallel to the slit (TE) and **b** perpendicular to the slit (TM). When the polarization is TE, a strong electrical field is generated just under the aperture. However, when the polarization is TM, a strong electrical field is generated around the edge of mask and under the mask. These phenomena due to the plasmon in the metal mask and diffraction

in the resist are different and depend on the slits' pitch. Such a pitch dependency is caused by the diffraction of light by a periodic slit pattern. Especially when the pitch of the slit is λ/n (λ: wavelength of exposed light; n: refractive index of resist), the diffracted light propagates on the surface of the mask, and it affects the transferred pattern. Therefore the masks must be designed with consideration of these results.

4.4.3 Dependency on Polarization

Figure 4.10 shows the results of simulating electric fields in the resist layer generated from the slit when the polarization of exposure light was parallel to the slit (TE) and perpendicular to the slit (TM). When the polarization is TE, a strong electrical field is generated just under the aperture. However, when the polarization is TM, a strong electrical field is generated around the edge of mask and under the mask. These phenomena are due to the plasmon in the metal mask and diffraction.

4.5 Application

As one application of NFOL, we made an ultraviolet second harmonic generation (SHG) wavelength conversion device. The periodically polled crystal of MgO–LiNbO3 described in Fig. 4.11 was put to practical use as a stable and efficient wavelength-converter device [19–21]. Applying the voltage between the periodical electrodes on a surface of crystal and other surfaces makes this device. To make the device for shorter wavelengths, the pitch of the poll's period must be shorter. Because the poll's domain is formed fatter than the electrodes, as shown in Fig. 4.12, the electrode must be sufficiently narrower than the desired pitches. For example, to make the device for a wavelength than 400 nm, the width of the periodic electrodes must be smaller than 300 nm. A conventional exposure device with g-line cannot achieve such a width, because its resolution is more than 500 nm. On the other hand, NFOL system can easily achieve such a resolution. Figure 4.13 shows the SEM image of electrodes on the MgO–LiNbO3 made by NFOL process. As shown in this image, 300-nm wide electrodes were formed. Using these electrodes, we could make the optical wavelength conversion device that changes 730 nm light into 365 nm light. Figure 4.14 shows the experimental results for second harmonic generation.

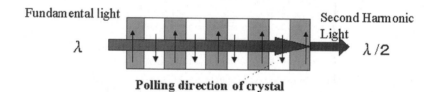

Fig. 4.11. The periodically polled crystal of MgO–LiNbO3 was put to practical use as a stable and efficient wavelength-converter device. To make the device for shorter wavelengths, the pitch of the poll's period must be shorter

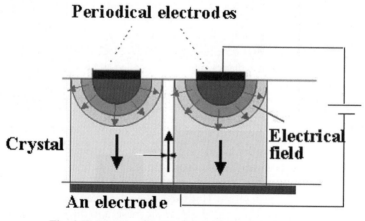

Fig. 4.12. Domain of poll is formed fatter than electrodes

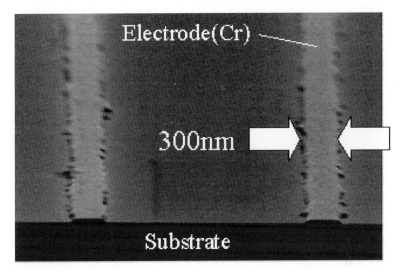

Fig. 4.13. SEM image of electrodes on the MgO–LiNbO₃ made by NFOL process. Electrodes 300-nm wide were formed

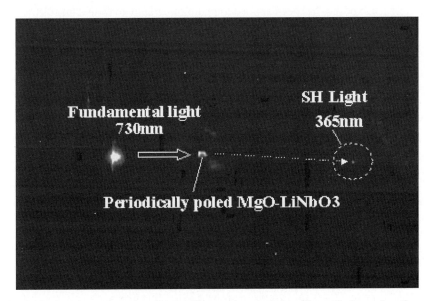

Fig. 4.14. Experimental results of second harmonic generation. The optical wavelength conversion device that changes 730-nm light into 365-nm light can be observed

NFOL is a low-cost, high-resolution method for fine processing. One important application is making a optical grating. Recently, using subwavelength grating as the refraction index controlled device and the optical sensing device without material design has become attractive. It is necessary, however, to take the structure for the grating at some cycles to exist together. This is especially true for the source of light of wavelength division multiplexing, though a diffraction grating structure (DFB, DBR) has been installed to control the wavelength-to-semiconductor laser used for the communication etc. Semiconductor laser types in which a diffraction grating is formed on a region along an active layer to select and stabilize an oscillating frequency—such as distributed Bragg reflector (DBR) lasers, distributed feedback (DFB) lasers and the like—have hitherto been provided in a variety of ways. Recently, in order to realize a blue light source in this type of semiconductor layer, there is an increasing demand for shortening the oscillating wavelength. To meet this demand for a shorter wavelength, a technique for fabricating a diffraction grating with a microcycle, as well as selection of semiconductor materials, becomes indispensable. If this diffraction grating is made into a higher-order diffraction grating, the cycle becomes relatively large and the processing consequently becomes easy. However, in the higher-order diffraction grating there is a reduction in the feedback light quantity due to spatially diffracted light, and on top of that, there is a need to control the line-and-space ratio with a high degree of accuracy. Therefore, in many cases, a first-order diffraction grating is preferred. In the case of a first-order diffraction grating, the linewidth of the grating pattern on the order of 100 nm or less is required. In widely used methods for fabricating a fine diffraction grating, two-beam interference exposure has been applied to photolithography to form a fine diffraction grating pattern. However, since in optical lithography the diffraction limit of light becomes the limit of resolution, it has been said that a linewidth of 100 nm (in terms of a diffraction grating cycle, 200 mm) is the limit of fining, even if an F_2 eximer laser of wavelength 248 nm is employed. Furthermore, if a resolution on the order of a nanometer greater than 100 nm is to be obtained, electron beam lithography or X-ray (particularly, synchrotron radiation) lithography must be employed. The e-beam lithography has the following advantages: basically a pattern on the order of a nanometer can be controlled and formed with a high degree of accuracy; the focal depth is considerably deep compared with optical systems; and direct wafer exposure (i.e., exposure of the resist directly by focused electron beam without a mask) is possible. However, e-beam lithography has the drawback that it is still far from mass production, because the throughput is low and it requires expensive devices that increase cost. Because it employs scanning exposure, e-beam lithography has a problem in that it is difficult to maintain a uniform linewidth with respect to a wide area. X-ray lithography is capable of enhancing resolution and precision by a factor of 10, compared with eximer laser exposure, even when 1 : 1 mask exposure is performed and even when a reflection-type image-forming X-ray optics system is employed. However, X-ray lithography's drawbacks are that mask fabrication is difficult and fabrication costs are high. In lithography employing an X-ray or e-beam, there is a need to develop specialized resists in accordance with the exposure methods, and there are still many problems associated with sensitivity, resolution, resistance to etching etc.

4.6 Summary

In this study, we presented a near-field optical photolithography experiment as an instance of nanofabrication using near-field light. We employed the bilayer resist and succeeded in forming a slit pattern with a line width of 130 nm and a depth of 550 nm. This suggests that fine patterns could be formed on a structure with a practical aspect ratio using near-field light exposure technologies. However, it is necessary to consider the structural dependency of the near-field pattern when we design near-field optical lithography process. Near-field optical photolithography technologies are considered suitable for creating gratings or nano dots, for instance, and are expected to be practically used as methods for fabricating the distributed feedback (DFB) laser, the distributed Bragg reflector (DBR) laser and communication optical devices.

References

[1] I.I. Smolyaninov, D.L. Mazzoni, C.C. Davis, Appl. Phys. Lett. **67**, 3859 (1995)
[2] S. Davy, M. Spajer, Appl. Phys. Lett. **69**, 3306 (1996)
[3] J. Massanell, N. Garcia, A. Zlatkin, Opt. Lett. **21**, 12 (1996)
[4] T. Ono, M. Esashi, Jpn. J. Appl. Phys. **37**, 6745 (1998)
[5] S. Tanaka, M. Esashi, Jpn. J. Appl. Phys. **37**, 6739 (1998)
[6] M.M. Alkaisi, R.J. Blaikie, S.J. McNab, R. Cheung, D.R.S. Cumming, Appl. Phys. Lett. **22**, 3560 (1999)
[7] J.G. Goodberlet, Appl. Phys. Lett. **76**, 667 (2000)
[8] M.M. Alkaisi, R.J. Blaikie, S.J. McNab, Adv. Mater. **13**, 1–11 (2001)
[9] R.J. Blaikie, M.M. Alkaisi, S.J. McNab, D.O.C. Melville, Int. J. Nanosci. **13**, 1–13 (2001)
[10] J.G. Goodberlet, Appl. Phys. Lett. **81**, 1315 (2002)
[11] W. Srituravanich, N. Fang, C. Sun, Q. Luo, X. Zhang, Nano Lett. **4**, 1088 (2004)
[12] M. Naya, I. Tsuruma, T. Tani, A. Mukai, Appl. Phys. Lett. **86**, 201113 (2005)
[13] T. Ito, M. Ogino, T. Yamada, Y. Inao, T. Yamaguchi, T. Mizutani, R. Kuroda, J. Photopoly. Sci. Technol. **18**, 435 (2005)
[14] T. Ito, T. Yamada, Y. Inao, T. Yamaguchi, N. Mizutani, R. Kuroda, Appl. Phys. Lett. **89**, 033113 (2006)
[15] T. Yatsui, Y. Nakajima, W. Nomura, M. Ohtsu, Appl. Phys. B **84**, 265 (2006)
[16] K. Sugita, N. Ueno, Prog. Polym. Sci. **17**, 319 (1992)
[17] A. Stainmann, Proc. SPIE **920**, 13 (1988)
[18] D.R. McKean, N.J. Klecak, L.A. Pederson, Proc. SPIE **1262**, 110 (1990)
[19] A. Harada, Y. Nihei, Appl. Phys. Lett. **69**, 2629 (1996)
[20] K. Mizuuchi, K. Yamamoto, M. Kato, Electron. Lett. **32**, 2091 (1996)
[21] S. Sonoda, I. Tsuruma, M. Hatori, Appl. Phys. Lett. **70**, 3078 (1997)

5

Nano-Optical Manipulation Using Resonant Radiation Force

T. Iida and H. Ishihara

5.1 Introduction

Techniques based on radiation force have enabled us to handle micron-sized objects in a noncontact manner. These techniques are now applied in a variety of research fields such as chemistry, biology, materials science and so on. On the other hand, extensive studies in the field of condensed matter photophysics have revealed that nanostructure electronic systems exhibit quantum mechanical characteristics in their optical responses to a resonant frequency region. These responses depend on the systems' properties with respect to size, shape, and so on. This article introduces the first investigation linking the above two elements, i.e., the radiation force and the world of quantum mechanics peculiar to nanoscale systems. The findings, obtained in a series of relevant studies, lead to novel techniques of optical manipulation that enables us to sort and select nano objects with particular quantum mechanical properties.

In the first section, we briefly review the conventional studies of optical manipulation including theoretical works and refer to the expected potentiality of optical manipulation using "resonant" radiation force.

5.1.1 Techniques Using Radiation Force

Optical manipulation is a technique for controlling the spatial configuration and the mechanical motions of small objects by using the "radiation force" induced by laser light (for review articles, see [1, 2]). Although the theoretical predictions on radiation pressure appeared in the 19th century and experimental demonstrations were performed in the early 20th century [3, 4], the technology-oriented studies appeared only after 1970, where Ashkin at AT&T Bell Laboratories theoretically demonstrated the possibility that atoms can be trapped using resonant radiation pressure [5]. With significant contributions of theoretical studies [6, 7] and technical improvements, researchers succeeded in cooling and trapping atoms using laser light [8–10]. In 1997, for establishing the methods of cooling and trapping atoms using laser light, S. Chu, C. Cohen-Tannoudji, and W.D. Phillips were awarded the Nobel Prize in Physics. Successively, E.A. Cornell, W. Ketterle, and C.E. Wieman achieved Bose–Einstein condensation in dilute gases of alkali atoms with a laser cooling technique and were awarded the Nobel Prize in Physics in 2001 [11, 12]. Another important work of Ashkin is the first experimental demonstration of the acceleration and trapping of micrometer-size dielectric spheres by using cw visible laser light [13]. The recent remarkable progress in the optical manipulation technique—showing its promise as a tool for handling small objects—started from his work. Particularly, in 1986, his group succeeded in trapping dielectric particles using a single focused laser beam, which has become the basis for what is called an optical tweezers for manipulating micron-sized objects [14].

Up to now, various fundamental and application studies on optical manipulation of micron-sized objects have been performed. One of the examples is the fabrication of photonic crystals by arranging micrometer-size polystyrene spheres by using the periodic optical traps [15]. In the field of biochemistry, the stretching of a DNA strand was demonstrated using the optical tweezers to trap the micrometer-size bead attached at the end of the strand [16]. Further, the application of laser-induced driving force to rotate a windmill-like particle was demonstrated; this will be useful for micro machining [17]. In addition, there are many demonstrations of manipulating micron-sized objects using radiation force [2, 18–20]. As one of the fundamental studies, a system was developed to precisely and instantaneously observe the potential energy of laser trapping as a function of three-dimensional position to analyze the radiation pressure acting on a single microparticle in a solution [21]. In these studies, high-power laser light of electronically nonresonant frequency is usually used to generate sufficiently strong force while not destroying the objects under study.

Recently, one of the major interests in this field is to manipulate smaller nanoscale objects. For metallic substances, the trapping of a single 36-nm diameter gold particle has been achieved [22] and a 40-nm diameter gold particle was employed as a probe of SNOM [23]. In addition, the analysis of the radiation pressure exerted on a gold particle with the diameter of 250 nm was demonstrated [24]. On the other hand, for nonmetallic particles, submicron-sized (diameter $\sim$220 nm) polystyrene latex particles with fluorescent dye have been trapped, gathered and fixed onto a substrate using a strongly focused laser beam [25]. However, if the particle is much smaller than the wavelength corresponding to the usual laser frequency region (e.g., diameter $<$100 nm), the induced polarization becomes very small; hence, the light-matter interaction becomes too weak to cause sufficiently strong force under the electronically nonresonant condition. Therefore, it is difficult to overcome the random force arising from the surrounding fluid medium that causes the Brownian motion at room temperature and to manipulate nano objects in the available frequency region.

5.1.2 Previous Theoretical Studies

For the given system consisting of matter and electromagnetic fields, we can calculate the radiation force on the matter in principle by substituting the appropriate response fields into the following equation, i.e., the equation of Lorentz force.

$$\mathbf{F} = \frac{d\mathbf{G}_m(t)}{dt} = \sum_i e_i \left[\mathbf{E}(\mathbf{r}_i, t) + \frac{1}{c} \mathbf{v}_i \times \mathbf{H}(\mathbf{r}_i, t) \right] \tag{5.1}$$

$$= \int d\mathbf{r} \left[\rho(\mathbf{r}, t)\mathbf{E}(\mathbf{r}, t) + \frac{1}{c}\mathbf{J}(\mathbf{r}, t) \times \mathbf{H}(\mathbf{r}, t) \right], \tag{5.2}$$

where $\mathbf{G}_m$ is the momentum of the center-of-mass of the matter system; c is the light velocity in vacuum; e_i and $\mathbf{v}_i$ are the electric charge of the ith particle and its velocity, respectively; and $\rho(\mathbf{r}, t)$, $\mathbf{J}(\mathbf{r}, t)$, $\mathbf{E}(\mathbf{r}, t)$, and $\mathbf{H}(\mathbf{r}, t)$ are the charge density, current density, electric field and magnetic field at position $\mathbf{r}$, respectively. For actual calculations, it is often preferred to perform the surface integration of the Maxwell stress tensor rather than the volume integration based on the Lorentz force. Using Maxwell's equations, we can recast (5.2) in a surface integration of the Maxwell stress tensor, namely,

$$\mathbf{F} = \frac{d}{dt} \left[\mathbf{G}_m(t) + \frac{1}{4\pi c} \int_V d\mathbf{r} \{ \mathbf{E}(\mathbf{r}, t) \times \mathbf{H}(\mathbf{r}, t) \} \right] = \int_S dS \bar{\mathbf{T}}(\mathbf{r}, t) \cdot \mathbf{n}(\mathbf{r}), \tag{5.3}$$

where $\mathbf{n}(\mathbf{r})$ is the surface normal unit vector pointing outwards, $\bar{\mathbf{T}} = [\mathbf{E}\mathbf{E} + \mathbf{H}\mathbf{H} - (1/2)(|\mathbf{E}|^2 + |\mathbf{H}|^2)\mathbf{I}]/4\pi$ is the Maxwell stress tensor, and $\mathbf{I}$ is the unit tensor (see Sect. 5.2.1 for details). As shown in (5.3), the Maxwell stress tensor integrated over a closed surface S is the rate of balance between the incoming and outgoing momentum flows in region V enclosed by surface S, which should be equal to the time derivative of the momenta of the total system of matter and radiation in this region. For steady state, the second term of the central part of (5.3), namely, the time variation of the field momentum, becomes zero when the time dependence of the fields is described as $\exp[-i\omega t]$. Therefore, the right-hand side is equivalent to the Lorentz force on the matter in a steady state, if we choose S as the surface of the matter.

On the basis of the above macroscopic framework, some important theoretical studies on radiation force have been conducted. In 1909, Debye rigorously derived the analytical expression of the radiation pressure for a steady state using the Mie scattering theory and the Maxwell stress tensor method when a spherical particle is irradiated by the plane wave propagating in the z-direction as

$$F_z = \frac{1}{8\pi} \left(\sigma_{scat}\left(1 - \overline{\cos\theta} \right) + \sigma_{abs} \right) I^{(i)}, \tag{5.4}$$

where $I^{(i)}$ is the intensity of the incident field; $\overline{\cos\theta}$ is the asymmetry parameter; and σ_{scat} and σ_{abs} are the scattering and the absorption cross-sections of light, respectively [26–28]. σ_{scat} and σ_{abs} are proportional to the momenta of the photons elastically scattered and absorbed by the matter per unit of time, respectively. Therefore, (5.4) indicates that the radiation pressure can be understood as a sum of "scattering force" and "absorbing force" arising from the momentum transfer from the light. On the other hand, if we consider an electrically neutral particle as a mass point that moves in electromagnetic fields, the Lorentz force on it can be expressed as

$$\mathbf{f} = (\mathbf{p} \cdot \nabla)\mathbf{E}^{(i)} + \frac{1}{c}\frac{d\mathbf{p}}{dt} \times \mathbf{B}^{(i)} \tag{5.5}$$

$$= \frac{1}{2}\alpha\nabla(\mathbf{E}^{(i)})^2 + \frac{1}{c}\alpha\frac{\partial}{\partial t}(\mathbf{E}^{(i)} \times \mathbf{B}^{(i)}), \tag{5.6}$$

where $\mathbf{p} = \alpha\mathbf{E}^{(i)}$ is the dipole moment of the Rayleigh particle; α is the polarizability; and $\mathbf{E}^{(i)}$ and $\mathbf{B}^{(i)}$ are the electric and magnetic fields of the incident light at the position of the particle, respectively [29]. It is considered that the first term of (5.6) corresponds to the force pushing objects to the focal point of the laser beam used in optical tweezers, as mentioned in Sect. 5.1.1, which is called "gradient force" (dipole force or reactive force) because the magnitude of the force is proportional to the gradient of the intensity of light (see Fig. 5.1). This type of force arises even if the matter is irradiated by the nonpropagating light. On the other hand, the second term of (5.6) is proportional to the time derivative of Poynting vector $\mathbf{E} \times \mathbf{H}$ of electromagnetic field. Thus, this term is considered as a force corresponding to (5.4).

Following the above theoretical studies, many demonstrations of the calculation of radiation force on macroscopic objects have been performed. For micron-sized objects, the calculation of net radiation force and torque for a spherical particle illuminated by a focused laser beam was demonstrated by means of the Maxwell stress tensor method [30]. Further, the numerical estimation of the force on a thin film in an evanescent field with a pure imaginary wave number was demonstrated [31]. The calculated gradient forces and experimental results were compared for the particle in strongly localized fields, and a good agreement was obtained between the calculated and experimental results [32]. In the ray optics regime, the analytical expressions of the radiation forces exerted on a micrometer-size dielectric sphere by a single-beam gradient laser trap were derived [33]. For nanoparticles, the manipulation using strongly localized evanescent fields near a metallic probe or a nanoaper-

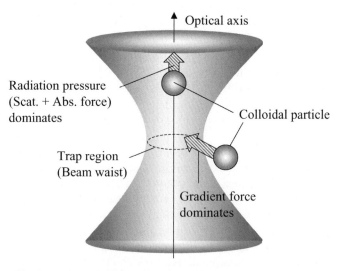

Fig. 5.1. Schematic illustration of the principle of optical tweezers

ture was proposed based on calculations performed using the finite difference time domain (FDTD) method [34] and the Maxwell stress tensor method [35, 36], where the steep gradient of the field intensity generates a strong gradient force. With the model calculation based on the discrete dipole approximation (DDA) [37], the properties of the radiation force on a metallic particle or a dielectric particle (without electronic resonance) were investigated for an object located in a plane wave field or an evanescent field from the dielectric substrate [38]. The same group also discussed the mechanical interaction between two particles without electronic resonance in the presence of a light field from the dielectric substrate [39]. They further proposed the optical trapping and manipulation of nano objects with an apertureless probe under the illumination of an evanescent field from the substrate [40]. On the other hand, other researchers calculated the radiation force on a dielectric Rayleigh particle irradiated by a focused Gaussian beam with the frequency near the electronic resonance [42], where they used the point dipole approximation, neglecting the spatial structure of the wavefunction of the particle's electronic state and assuming classical polarizability with a single Lorentz-type resonance level. Since the targets of these previous studies are macroscopic objects, their optical properties are compressed into the macroscopic parameters such as polarizability, dielectric function and refractive index.

With regard to the study based on a microscopic theory, analytical expressions of the resonant radiation force on an atom are derived [6, 7]. The radiation force on a two-level atom at rest (its moving distance is much smaller than the light wavelength during the excitation and relaxation) in vacuum can be derived from the Heisenberg equation in the semiclassical limit as

$$\mathbf{F} = \mathbf{F}_{\text{dissip}} + \mathbf{F}_{\text{react}}, \tag{5.7}$$

$$\mathbf{F}_{\text{dissip}} = \hbar \mathbf{k}_{\text{L}} \frac{\Gamma}{2} \frac{\Omega_1^2/2}{(\omega_{\text{A}} - \omega_{\text{L}})^2 + (\Gamma^2/4) + (\Omega_1^2/2)} = \hbar \mathbf{k}_{\text{L}} \left\langle \frac{dN}{dt} \right\rangle_{\text{st}}, \tag{5.8}$$

$$\mathbf{F}_{\text{react}} = \frac{\hbar(\omega_{\text{A}} - \omega_{\text{L}})}{4} \frac{\nabla \Omega_1^2}{(\omega_{\text{A}} - \omega_{\text{L}})^2 + (\Gamma^2/4) + (\Omega_1^2/2)}, \tag{5.9}$$

where ω_{A} is the resonance frequency of an atom; ω_{L} and $\mathbf{k}_{\text{L}}$ are the frequency and the wave vector of an incident laser light, respectively; Ω_1 is the Rabi frequency proportional to the amplitude of the incident field $|\mathbf{E}^{(i)}|$ (the strength of the interaction between the atomic dipole and incident field); Γ is the natural width (full width at half maximum); and $\langle dN/dt \rangle_{\text{st}}$ is the mean number of photons absorbed per unit time in the steady state. The semiclassical limit implies that the atom and the vacuum field are treated quantum mechanically, while the incident laser field is treated classically. The dissipative force $\mathbf{F}_{\text{dissip}}$ in (5.8) corresponds to the scattering force proportional to the time variation of the photon momentum $\hbar \mathbf{k}_{\text{L}}$ transferred to the atom. The reactive force $\mathbf{F}_{\text{react}}$ in (5.9) corresponds to the gradient force because its magnitude is proportional to the gradient of the incident intensity. We can see that both types of forces are enhanced when the atom is irradiated by the laser light whose frequency ω_{L} is near ω_{A}. For low-intensity laser irradiation, $\mathbf{F}_{\text{dissip}}$ is proportional to $|\mathbf{E}^{(i)}|^2$. At high intensity, $\mathbf{F}_{\text{dissip}}$ is saturated and close to the value of $\hbar k_L \Gamma/2$, which is 10^5 times greater than the gravitational force in the case of Na atom ($\Gamma^{-1} = 16 \times 10^{-9}$ [sec]). On the other hand, $\mathbf{F}_{react}$ does not saturate even if the intensity infinitely increases. In this calculation, the spatial extension of the electronic wavefunction on the atom is not explicitly given, and the long-wavelength approximation is considered, where the spatial variation of the light field is also neglected.

5.1.3 Potentiality in Using Resonant Radiation Force in a Nanoscale Regime

In spite of extensive studies on optical manipulation for two contrasting regimes, i.e., the micron-sized regime and the atomic regime, the study of the nanoscale regime has been fairly limited, as explained earlier. If we consider the optical manipulation of nano objects from the viewpoint of application of optical tweezers, the use of nonresonant light would be a serious limitation because the induced polarization of nano objects is too small to yield sufficiently strong force. However, if we use electronically resonant light, the induced polarization exhibits a great enhancement in general due to the resonance effect that causes a sufficiently strong enhancement of the scattering and absorbing forces or the gradient force even on nano objects as in the atom cooling technique, whereas the electronic states of nanostructures appear too complicated for the motion control using this technique. Although resonant excitation usually causes a heating problem in condensed matter because of the nonradiative decay of the excitation, we can carefully choose the conditions under which this problem is not very serious according to the target. If the electronic systems confined in the objects have high coherence, the radiation-matter coupling becomes large; hence, a large portion of excitation can decay as radiation faster than nonradiative decay. In such a case, the induced force is dominated by the coherent scattering with less absorption. Thus, depending on the conditions of the target, it will be possible to consider the interesting manipulation of nano objects.

A more important aspect in the use of resonant light is that it directly connects the radiation force with the quantum mechanical individuality of nanostructures. This is quite a new aspect that is not included in the optical manipulation of micron-sized objects or atomic systems. For example, the electronic systems confined in nanostructures exhibit their individual spectral structures due to the quantum confinement effect depending on their specific sizes, and the peak positions in the force spectra are sensitive to nanoscale-size changes. Therefore, by tuning the frequency of the laser beam, it will be possible to mechanically separate the objects of a particular size. Furthermore, the quantum mechanical individualities generally appear not only through the size dependence of the target, but also through the other dependences on various parameters such as shape, chirality and coherence. Thus, the particular properties of objects can be sorted and selected by using resonant light through the quantum mechanical resonance condition depending on such characteristics of individual nano objects. The possibility of such a new type of optical manipulation can be demonstrated through the study of the radiation-matter mechanical interaction under the resonance condition based on the theory. This allows us to explicitly treat spatial structures of electronic (elementary excitation) states and their self-consistent motion with the radiation.

The purpose of this article is to present the results of our recent studies on optical manipulation using resonant radiation force. The organization of this article is as follows: Section 5.2 explains the theoretical bases of our work and general properties of the radiation force in the presence of the electronically resonant light. In this part, we discuss the radiation force for different-sized regimes on the basis of a unified expression applicable from the atomic regime to the macroscopic regime. This discussion clarifies the relations between different expressions obtained so far with approximations peculiar to the respective size regimes. In Sect. 5.3, we demonstrate the numerical calculations of the resonant radiation force on a single semiconductor nanoparticle to discuss the feasibility of optical manipulation of nano objects, where we assume that excitonic center-of-mass motions are confined in the nano object. After analyzing the contributions of several types of forces, we propose a new type of optical manipulation where we explicitly use a resonant effect of the confined electronic states. In Sect. 5.4, we demonstrate the calculations considering an experiment performed under cryogenic conditions in superfluid He^4. Under these conditions, the line width of excitons is extremely narrow and the best performance of resonant optical manipulation can be demonstrated. In this part, we consider the finite line width of the laser beam and Brownian motion of the objects by the random force arising from the surrounding medium. In a recent experiment on the optical transport of nanoparticles, this possibility was verified, where we successfully manipulated selected nanoparticle by using resonant radiation force in superfluid He^4. We present the results of this experiment in Sect. 5.5. In the final Sect. 5.6, we present summary and future prospects.

5.2 Theoretical Bases

One of the interesting aspects of nano systems is the remarkable dependence of their resonant optical response on structural parameters such as size, shape and arrangement. This feature arises from the fact that the wavefunctions of elementary excitations (e.g., excitons) are coherent over the whole volume of the object. Since the coherent length of excitons is considerably long in general if the sample quality is sufficiently high, this quantum effect appears for a wide range of object size in the nanoscale regime. Therefore, the interplay between the spatial structures of the radiation field and those of the matter wavefunction plays an essential role, which appears in the peculiar dependence of optical response on size, shape and so on [43–45]. For studying such features of optical response, the microscopic nonlocal response theory [45–47] is useful. In this theory, the motions of the source polarizations and electromagnetic field are determined self-consistently where the microscopic spatial distributions of the electromagnetic field and the wavefunction of quantum mechanical states of matter are explicitly considered.

In order to reveal the features of resonant radiation force peculiar to the nano regime, we need to include the above aspect of microscopic optical response. In the present study, we substitute the charge density, current density and electromagnetic fields (ρ, $\mathbf{j}$, $\mathbf{E}$ and $\mathbf{H}$) determined by the nonlocal response theory into in the Lorentz force (5.2) or in the Maxwell's stress tensor (5.3) to calculate the radiation force reflecting the quantum mechanical structure of the target objects. Using this method, we can obtain the analytical expressions of the resonant radiation force exerted on the finite-sized system "from atom to macroscopic object". From this result, we can understand not only the characteristic features of the radiation force in the nano regime but also the relation between previously obtained results for macroscopic and atomic regimes. In the following calculations, we assume the input laser intensity as being within the linear response regime and electrically neutral and nonmagnetic objects as targets of manipulation.

5.2.1 Lorentz Force and Maxwell Stress Tensor

We start from the set of N equations of the Lorentz force for the system consisting of N point charges e_i $(i = 1, \ldots, N)$ under electromagnetic fields in free space (vacuum) as

$$m_i \frac{\mathrm{d}^2 \mathbf{r}_i(t)}{\mathrm{d}t^2} = e_i \mathbf{E}\big(\mathbf{r}_i(t), t\big) + \frac{e_i}{c} \dot{\mathbf{r}}_i(t) \times \mathbf{H}\big(\mathbf{r}_i(t), t\big), \tag{5.10}$$

where m_i is the mass of each charge; $\mathbf{E}(\mathbf{r}_i(t), t)$ and $\mathbf{H}(\mathbf{r}_i(t), t)$ are the electric and magnetic fields at the position of the ith charge $\mathbf{r}_i(t)$ at time t, respectively [48]. In (5.10), the summation over all N charges leads to

$$\frac{\mathrm{d}\mathbf{G}_\mathrm{m}(t)}{\mathrm{d}t} = \int_V \mathrm{d}\mathbf{r} \left[\rho(\mathbf{r}, t)\mathbf{E}(\mathbf{r}, t) + \frac{\mathbf{J}(\mathbf{r}, t)}{c} \times \mathbf{H}(\mathbf{r}, t) \right], \tag{5.11}$$

where we rewrite the summation of the momenta of all charges as $\mathbf{G}_\mathrm{m}(t) = \sum_i m_i \dot{\mathbf{r}}_i(t)$ and use the definitions

$$\sum_{i=1}^{N} e_i \dot{\mathbf{r}}_i(t) \delta^3\big(\mathbf{r} - \mathbf{r}_i(t)\big) = \mathbf{J}(\mathbf{r}, t), \tag{5.12}$$

$$\sum_{i=1}^{N} e_i \delta^3\big(\mathbf{r} - \mathbf{r}_i(t)\big) = \rho(\mathbf{r}, t). \tag{5.13}$$

Here, $\mathbf{J}(\mathbf{r}, t)$ and $\rho(\mathbf{r}, t)$ are the current density and the charge density at position $\mathbf{r}$. This expresses the integral form of the Lorentz force on the entire matter system. Substituting the response fields appropriate for the given system consisting of matter and electromagnetic fields into (5.11), we can calculate the radiation force on the matter regardless of whether the fields are macroscopic or microscopic in principle.

For practical calculations, using the Maxwell stress tensor method based on surface integration is often preferred rather than using the Lorentz force based on volume integration. By using the Maxwell's equations, we can recast (5.11) as

$$\mathbf{F} = \frac{\mathrm{d}}{\mathrm{d}t}\big[\mathbf{G}_\mathrm{m}(t) + \mathbf{G}_\mathrm{f}(t)\big] = \int_V \mathrm{d}\mathbf{r} \, \mathrm{div}\bar{\mathbf{T}}(\mathbf{r}, t) = \int_S \mathrm{d}S \bar{\mathbf{T}}(\mathbf{r}, t) \cdot \mathbf{n}(\mathbf{r}), \tag{5.14}$$

where $\mathbf{F}$ is the stress on the region enclosed by surface S, $\overline{\mathbf{T}}$ is the Maxwell's stress tensor defined as $[\mathbf{EE} + \mathbf{HH} - (1/2)(|\mathbf{E}|^2 + |\mathbf{H}|^2)\mathbf{I}]/4\pi$ [48], the field momentum $\mathbf{G}_f(t)$ represents $(1/4\pi c)\int_V d\mathbf{r}\mathbf{S}(\mathbf{r}, t)$. $\mathbf{n}(\mathbf{r})$ is the surface normal unit vector pointing outwards, and $\mathbf{I}$ is a unit tensor.

From (5.14), we can understand that the Maxwell stress tensor integrated over a closed surface S is the rate of the balance between the incoming and outgoing momentum flows in this region V, which should be equal to the time derivative of the momenta of matter and field in this region. If we consider the time harmonic electromagnetic fields as $\mathbf{E}(\mathbf{r}, t) = \mathbf{E}(\mathbf{r}, \omega)\exp[-i\omega t]$, $\mathbf{H}(\mathbf{r}, t) = \mathbf{H}(\mathbf{r}, \omega)\exp[-i\omega t]$, the time average of time derivative of Poynting vector becomes

$$
\begin{aligned}
\left\langle \frac{d}{dt}(\mathbf{E}(\mathbf{r}, t) \times \mathbf{H}(\mathbf{r}, t)) \right\rangle &= \frac{1}{2}\text{Re}\left[\frac{d}{dt}(\mathbf{E}(\mathbf{r}, t) \times \mathbf{H}(\mathbf{r}, t))^* \right] \\
&= \frac{1}{2}\text{Re}\left[-i\omega\mathbf{E}(\mathbf{r}, \omega)\exp[-i\omega t] \times \mathbf{H}(\mathbf{r}, \omega)^* \exp[i\omega t] \right. \\
&\qquad \left. + i\omega\mathbf{E}(\mathbf{r}, \omega)\exp[-i\omega t] \times \mathbf{H}(\mathbf{r}, \omega)^* \exp[i\omega t] \right] \\
&= 0,
\end{aligned}
\tag{5.15}
$$

where $\langle \rangle$ indicates time average. Thus, for the steady state, the time variation of the field momentum $\langle d\mathbf{G}_f(t)/dt \rangle$ becomes zero, and (5.14) can be rewritten as

$$
\langle \mathbf{F} \rangle (\omega) = \left\langle \frac{d}{dt}\mathbf{G}_m(t) \right\rangle = \left\langle \int_V d\mathbf{r}\, \text{div}\overline{\mathbf{T}}(\mathbf{r}, t) \right\rangle = \left\langle \int_S d\mathbf{S}\overline{\mathbf{T}}(\mathbf{r}, t) \cdot \mathbf{n}(\mathbf{r}) \right\rangle.
\tag{5.16}
$$

Therefore, the surface integration of the Maxwell stress tensor is equivalent to the Lorentz force on the matter in the steady state, if we choose S as the surface that encloses the matter.

5.2.2 Microscopic Response Field

For a unified description of the optical response from the microscopic to the macroscopic system, we should start from the nonlocal expression for the induced polarization. That is, the resonant part of the linear polarization should be

$$
\mathbf{P}_{\text{res}}(\mathbf{r}, \omega) = \int_V d\mathbf{r}'\chi(\mathbf{r}, \mathbf{r}'; \omega)\mathbf{E}(\mathbf{r}', \omega),
\tag{5.17}
$$

where $\chi(\mathbf{r}, \mathbf{r}'; \omega)$ is the linear susceptibility. According to the Kubo formula, the microscopic susceptibility is generally described as a separable form with respect to the microscopic positions $\mathbf{r}$ and $\mathbf{r}'$ as

$$
\chi(\mathbf{r}, \mathbf{r}'; \omega) = \sum_\lambda \left[\frac{\rho_{0\lambda}(\mathbf{r})\rho_{\lambda 0}(\mathbf{r}')}{E_\lambda - \hbar\omega - i\gamma} + \frac{\rho_{\lambda 0}(\mathbf{r})\rho_{0\lambda}(\mathbf{r}')}{E_\lambda + \hbar\omega + i\gamma} \right]
\tag{5.18}
$$

with the matrix element of the transition dipole density from the νth electronic level to the μth one as

$$\boldsymbol{\rho}_{\mu\nu}(\mathbf{r}) = \boldsymbol{\rho}_{\nu\mu}^{*}(\mathbf{r}) = \langle\mu|\hat{\mathbf{P}}(\mathbf{r})|\nu\rangle. \tag{5.19}$$

In these expressions, $\{E_{\lambda}\}$ are the eigenenergies of the eigenstates $\{|\lambda\rangle\}$ of the unperturbed system. In (5.18), γ is the nonradiative damping parameter, where we treat nonradiative scattering mechanisms phenomenologically and γ is taken to be a positive finite value.

For the time harmonic electromagnetic field, the following equation for the electric field can be derived from the Maxwell equation:

$$\nabla \times \nabla \times \mathbf{E}(\mathbf{r}, \omega) - q^{2}\mathbf{E}(\mathbf{r}, \omega) = 4\pi q^{2}\mathbf{P}(\mathbf{r}, \omega), \tag{5.20}$$

where $q = \omega/c$ is the wave number of the incident light in vacuum. The Maxwell field $\mathbf{E}(\mathbf{r}, \omega)$ that satisfies this equation generally includes the longitudinal electric field $\mathbf{E}_{L}$ as

$$\mathbf{E}(\mathbf{r}, \omega) = \mathbf{E}_{L}(\mathbf{r}, \omega) + \mathbf{E}_{T}(\mathbf{r}, \omega), \tag{5.21}$$

$$\mathbf{E}_{L}(\mathbf{r}, \omega) = \nabla\nabla \cdot \int_{V} d\mathbf{r}' \frac{1}{|\mathbf{r} - \mathbf{r}'|}\mathbf{P}(\mathbf{r}', \omega), \tag{5.22}$$

where $\mathbf{E}_{T}(\mathbf{r}, \omega)$ is the transverse part of $\mathbf{E}(\mathbf{r}, \omega)$. Here, we divide the induced polarization $\mathbf{P}$ into the resonant part $\mathbf{P}_{res}$ and the nonresonant part $\mathbf{P}_{nonres}$, and assume that the latter can be described by a local background susceptibility χ_{b} including the second term (nonresonant part) in (5.18). For $\mathbf{r}$ inside or outside the background dielectric, the nonresonant polarization has the form

$$\mathbf{P}_{nonres}(\mathbf{r}, \omega) = \chi_{b}\Theta(\mathbf{r})\mathbf{E}(\mathbf{r}, \omega),$$

$$\Theta(\mathbf{r}) = \begin{cases} 1 & (\mathbf{r} \in \text{inside}), \\ 0 & (\mathbf{r} \in \text{outside}). \end{cases} \tag{5.23}$$

By using (5.23), the Maxwell equation (5.20) can be rewritten as

$$\nabla \times \nabla \times \mathbf{E}(\mathbf{r}, \omega) - q^{2}[1 + 4\pi\chi_{b}\Theta(\mathbf{r})]\mathbf{E}(\mathbf{r}, \omega) = 4\pi q^{2}\mathbf{P}_{res}(\mathbf{r}, \omega). \tag{5.24}$$

For this equation, we can define the renormalized dyadic Green function $\mathbf{G}^{b}(\mathbf{r}, \mathbf{r}', \omega)$ described by the following equation

$$\nabla \times \nabla \times \mathbf{G}^{b}(\mathbf{r}, \mathbf{r}', \omega) - q^{2}[1 + 4\pi\chi_{b}\Theta(\mathbf{r})]\mathbf{G}^{b}(\mathbf{r}, \mathbf{r}', \omega) = 4\pi q^{2}\mathbf{I}\delta(\mathbf{r} - \mathbf{r}'), \tag{5.25}$$

where $\mathbf{I}$ is a unit tensor. For simple geometries of the background medium, such as a planar medium or a spherical medium, analytical solutions of (5.25) are known [49]. In terms of this renormalized Green function, we can describe the solution of (5.24) as

$$\mathbf{E}(\mathbf{r}, \omega) = \mathbf{E}^{b}(\mathbf{r}, \omega) + \int_{V} d\mathbf{r}'\mathbf{G}^{b}(\mathbf{r}, \mathbf{r}', \omega) \cdot \mathbf{P}_{res}(\mathbf{r}', \omega), \tag{5.26}$$

where $\mathbf{E}^b(\mathbf{r}, \omega)$ represents the homogeneous solution, i.e., the field in the presence of the background dielectric alone, which may contain longitudinal components as well as transverse ones, in contrast to the case in vacuum. Since the polarization is induced by the electric field, as shown in (5.17), and the former drives the latter following (5.26), the response field should be determined self-consistently by solving these equations simultaneously. For this purpose, the microscopic nonlocal theory by Cho provides a systematic methodology [45–47]. By defining the quantity representing the interaction between the induced polarization and the response Maxwell field over a resonant denominator as

$$X_\lambda(\omega) \equiv \frac{1}{E_\lambda - \hbar\omega - i\gamma} \int_V d\mathbf{r}\boldsymbol{\rho}_{\lambda 0}(\mathbf{r}) \cdot \mathbf{E}(\mathbf{r}, \omega), \qquad (5.27)$$

we can expand the resonant polarization $\mathbf{P}_{res}$ by using the basis of the matter system as

$$\mathbf{P}_{res}(\mathbf{r}, \omega) = \sum_\lambda X_\lambda(\omega)\boldsymbol{\rho}_{\lambda 0}^*(\mathbf{r}), \qquad (5.28)$$

from (5.17) and (5.18), and obtain

$$\mathbf{E}(\mathbf{r}, \omega) = \mathbf{E}^b(\mathbf{r}, \omega) + \sum_\lambda X_\lambda(\omega) \int_V d\mathbf{r}' \mathbf{G}^b(\mathbf{r}, \mathbf{r}', \omega) \cdot \boldsymbol{\rho}_{\lambda 0}^*(\mathbf{r}'). \qquad (5.29)$$

The integral of the inner products of $\boldsymbol{\rho}_{\lambda 0}$ and (5.29) over volume V lead to the linear simultaneous equations of X_λ as

$$(E_\lambda - \hbar\omega - i\gamma)X_\lambda(\omega) + \sum_{\lambda'} A_{\lambda\lambda'}(\omega)X_{\lambda'}(\omega) = X_\lambda^{(b)}(\omega), \qquad (5.30)$$

$$A_{\lambda 0;0\lambda'}(\omega) \equiv - \iint_V d\mathbf{r}\,d\mathbf{r}' \boldsymbol{\rho}_{\lambda 0}(\mathbf{r}) \cdot \mathbf{G}^b(\mathbf{r}, \mathbf{r}', \omega)\boldsymbol{\rho}_{\lambda' 0}^*(\mathbf{r}'), \qquad (5.31)$$

$$X_\lambda^{(b)}(\omega) \equiv \int_V d\mathbf{r}\boldsymbol{\rho}_{\lambda 0}(\mathbf{r}) \cdot \mathbf{E}^b(\mathbf{r}, \omega), \qquad (5.32)$$

where $A_{\lambda 0;0\lambda'}(\omega)$ is the interaction between polarizations via the electromagnetic field. This quantity provides the complex self-energy corresponding to the radiative correction including the shift and width. $X_\lambda^{(b)}(\omega)$ is the interaction between the induced polarization and the incident field modified by the background dielectric. We can rewrite these linear equations into a matrix form as

$$\bar{\mathbf{S}}\mathbf{X} = \mathbf{X}^{(b)},$$

where

$$\begin{aligned}
\bar{\mathbf{S}} &= [(E_\lambda - \hbar\omega - i\delta)\delta_{\lambda\lambda'} + A_{\lambda 0;0\lambda'}(\omega)], \\
\mathbf{X} &= [X_\lambda]^\mathrm{T}, \\
\mathbf{X}^{(b)} &= [X_\lambda^{(b)}]^\mathrm{T}.
\end{aligned} \qquad (5.33)$$

In these expressions, ^{T} implies a vertical vector. The solution vector $\mathbf{X}$ can be obtained by multiplying both sides with the inverse matrix $\mathbf{S}^{-1}$ as

$$\mathbf{X} = \bar{\mathbf{S}}^{-1}\mathbf{X}^{(b)}. \tag{5.34}$$

This equation indicates that the resonance poles of $\mathbf{X}$ correspond to the complex solutions of $\det \bar{\mathbf{S}}(\omega) = 0$. These solutions are the nontrivial solutions to (5.33) in the absence of an incident field ($\mathbf{X}^{(b)} = 0$) and provide the complex eigenmodes of the light-matter coupled system. We can call these modes self-sustaining modes because they correspond to nonzero amplitudes of polarizations and fields supporting one another to form eigenmodes. Finally, substituting these solutions $\mathbf{X}$ into (5.29), we can uniquely determine the microscopic self-consistent response fields.

5.2.3 Derivation of General Expressions

Generally, the charge density can be described as $\rho = \rho_{\text{true}} + \rho_{\text{pol}}$ for the condensed matter considered as assembly of charged particles, where ρ_{true} is the true charge density and $\rho_{\text{pol}} = -\nabla \cdot \mathbf{P}$ is the polarization charge density. For electrically neutral matter, ρ_{true} vanishes and only ρ_{pol} remains. Therefore, from the continuity equation $\mathbf{J} = \partial \mathbf{P}/\partial t$, the Lorentz force (5.11) on the neutral object can be described as

$$\frac{d\mathbf{G}_{\text{m}}(t)}{dt} = \int_V d\mathbf{r}\left[-(\nabla \cdot \mathbf{P}(\mathbf{r}, t))\mathbf{E}(\mathbf{r}, t) + \frac{1}{c}\frac{\partial \mathbf{P}(\mathbf{r}, t)}{\partial t} \times \mathbf{H}(\mathbf{r}, t)\right]. \tag{5.35}$$

With the relations of $\mathbf{H}(\mathbf{r}, \omega) = c\nabla \times \mathbf{E}(\mathbf{r}, \omega)/i\omega$ from the Maxwell equations and $\mathbf{J}(\mathbf{r}, \omega) = -i\omega\mathbf{P}(\mathbf{r}, \omega)$ from the continuity equation for the time harmonic electromagnetic fields, we can rewrite the time averaged force as

$$\langle \mathbf{F}\rangle(\omega) = \left\langle \frac{d\mathbf{G}_{\text{m}}(t)}{dt}\right\rangle$$

$$= \frac{1}{2}\text{Re}\left[\int_V d\mathbf{r}\left[-(\nabla \cdot \mathbf{P}(\mathbf{r}, \omega))\mathbf{E}(\mathbf{r}, \omega)^* + \mathbf{P}(\mathbf{r}, \omega) \times \nabla \times \mathbf{E}(\mathbf{r}, \omega)^*\right]\right] \tag{5.36}$$

in the steady state. Since the polarization at the surface vanishes as $\mathbf{P}(\mathbf{r} = \text{surface}) = 0$, with integration by parts, the first term on the right-hand side of (5.36) becomes

$$\int_V d\mathbf{r}(\nabla \cdot \mathbf{P}(\mathbf{r}, \omega))\mathbf{E}(\mathbf{r}, \omega)^* = -\int_V d\mathbf{r}(\mathbf{P}(\mathbf{r}, \omega) \cdot \nabla)\mathbf{E}(\mathbf{r}, \omega)^*. \tag{5.37}$$

Using the vector identity

$$\mathbf{a} \times \nabla \times \mathbf{c} = (\nabla \mathbf{c}) \cdot \mathbf{a} - (\mathbf{a} \cdot \nabla)\mathbf{c}, \tag{5.38}$$

we can rewrite the second term on the right-hand side of (5.36) as

$$\int_V d\mathbf{r}\mathbf{P}(\mathbf{r}, \omega) \times \nabla \times \mathbf{E}(\mathbf{r}, \omega)^*$$

$$= \int_V d\mathbf{r}\left[(\nabla \mathbf{E}(\mathbf{r}, \omega)^*) \cdot \mathbf{P}(\mathbf{r}, \omega) - (\mathbf{P}(\mathbf{r}, \omega) \cdot \nabla)\mathbf{E}(\mathbf{r}, \omega)^*\right]. \tag{5.39}$$

Substituting (5.37) and (5.39) into (5.36), we obtain a general expression of the radiation force as

$$\langle \mathbf{F} \rangle (\omega) = \left\langle \frac{d\mathbf{G}_m(t)}{dt} \right\rangle = \frac{1}{2} \text{Re} \left[\int_V dr \big(\nabla \mathbf{E}(\mathbf{r}, \omega)^* \big) \cdot \mathbf{P}(\mathbf{r}, \omega) \right]. \tag{5.40}$$

If we use the microscopic response field (5.29) and the induced polarization $\mathbf{P} = \mathbf{P}_{\text{nonres}} + \mathbf{P}_{\text{res}}$ described by (5.23) and (5.28), we can rewrite the radiation force in the presence of electronically resonant light as

$$\langle \mathbf{F} \rangle (\omega) = \frac{1}{2} \text{Re} \left[\int_V dr \big(\nabla \mathbf{E}(\mathbf{r}, \omega)^* \big) \cdot \mathbf{P}_{\text{nonres}}(\mathbf{r}, \omega) \right.$$
$$\left. + \int_V dr \big(\nabla \mathbf{E}(\mathbf{r}, \omega)^* \big) \cdot \mathbf{P}_{\text{res}}(\mathbf{r}, \omega) \right]$$
$$= \langle \mathbf{F}_{\text{nonres}} \rangle (\omega) + \langle \mathbf{F}_{\text{res}} \rangle (\omega), \tag{5.41}$$

$$\langle \mathbf{F}_{\text{nonres}} \rangle (\omega) = \frac{1}{2} \text{Re} \left[\chi_b \int_V dr \big(\nabla \mathbf{E}(\mathbf{r}, \omega)^* \big) \cdot \mathbf{E}(\mathbf{r}, \omega) \right], \tag{5.42}$$

$$\langle \mathbf{F}_{\text{res}} \rangle (\omega) = \frac{1}{2} \text{Re} \left[\sum_\lambda X_\lambda(\omega) \int_V dr \big(\nabla \mathbf{E}(\mathbf{r}, \omega)^* \big) \cdot \boldsymbol{\rho}_{\lambda 0}(\mathbf{r})^* \right] \tag{5.43}$$

which can be applied to any system from an atom to a macroscopic object, in principle. We can explicitly rewrite (5.42) to

$$\langle \mathbf{F}_{\text{nonres}} \rangle (\omega) = \frac{1}{2} \text{Re} \left[\chi_b \int_V dr \big(\nabla \mathbf{E}(\mathbf{r}, \omega)^* \big) \cdot \mathbf{E}(\mathbf{r}, \omega) \right]$$
$$= \frac{1}{2} \text{Re} \left[\int_V dr \big(\nabla \mathbf{E}^b(\mathbf{r}, \omega)^* \big) \cdot \mathbf{P}_{b-b}(\mathbf{r}, \omega) \right.$$
$$+ \int_V dr \big(\nabla \mathbf{E}^b(\mathbf{r}, \omega)^* \big) \cdot \mathbf{P}_{b-\text{res}}(\mathbf{r}, \omega)$$
$$+ \sum_{\lambda'} X_{\lambda'}(\omega)^* \int_V dr \left\{ \int_V dr' \big(\nabla \mathbf{G}^b(\mathbf{r}, \mathbf{r}', \omega)^* \big) \cdot \boldsymbol{\rho}_{\lambda' 0}(\mathbf{r}') \right\}$$
$$\times \mathbf{P}_{b-b}(\mathbf{r}, \omega),$$
$$+ \sum_{\lambda'} X_{\lambda'}(\omega)^* \int_V dr \left\{ \int_V dr' \big(\nabla \mathbf{G}^b(\mathbf{r}, \mathbf{r}', \omega)^* \big) \cdot \boldsymbol{\rho}_{\lambda' 0}(\mathbf{r}') \right\}$$
$$\times \mathbf{P}_{b-\text{res}}(\mathbf{r}, \omega) \Big], \tag{5.44}$$

$$\mathbf{P}_{b-b}(\mathbf{r}, \omega) = \chi_b \mathbf{E}^b(\mathbf{r}, \omega), \tag{5.45}$$

$$\mathbf{P}_{b-\text{res}}(\mathbf{r}, \omega) = \chi_b \sum_\lambda X_\lambda(\omega) \int_V dr' \mathbf{G}^b(\mathbf{r}, \mathbf{r}', \omega) \rho_{\lambda 0}^*(\mathbf{r}'), \tag{5.46}$$

where $\mathbf{P}_{b-b}(\mathbf{r}, \omega)$ is the nonresonant polarization induced by the incident electric field in the presence of the background dielectric, and $\mathbf{P}_{b-\text{res}}(\mathbf{r}, \omega)$ is the nonresonant polarization induced by the resonantly scattered field; further, (5.43) can be rewritten as

$$\langle \mathbf{F}_{\text{res}}\rangle(\omega) = \frac{1}{2}\text{Re}\left[\sum_\lambda X_\lambda(\omega)\int_V d\mathbf{r}\big(\nabla \mathbf{E}(\mathbf{r},\omega)^*\big)\cdot \boldsymbol{\rho}_{\lambda 0}(\mathbf{r})^*\right]$$

$$= \frac{1}{2}\text{Re}\left[\sum_\lambda X_\lambda(\omega)\int_V d\mathbf{r}\big(\nabla \mathbf{E}^{\text{b}}(\mathbf{r},\omega)^*\big)\cdot \boldsymbol{\rho}_{\lambda 0}(\mathbf{r})^*\right.$$

$$+ \sum_\lambda \sum_{\lambda'} X_\lambda(\omega) X_{\lambda'}(\omega)^*$$

$$\left. \times \int_V d\mathbf{r}\left\{\int_V d\mathbf{r}'\big(\nabla \mathbf{G}^{\text{b}}(\mathbf{r},\mathbf{r}',\omega)^*\big)\cdot \boldsymbol{\rho}_{\lambda' 0}(\mathbf{r}')\right\}\cdot \boldsymbol{\rho}_{\lambda 0}(\mathbf{r})^*\right].$$

$$(5.47)$$

The Mie resonance frequencies $\{\omega_\zeta\}$ are determined by the macroscopic Maxwell equations with only the background dielectric and will exceed the usual laser frequency region if the target object is much smaller than the light wavelength. In such a case, we can usually neglect $\langle \mathbf{F}_{\text{nonres}}\rangle(\omega)$ in (5.42) as a force due to the nonresonant polarization; hence, (5.41) can be approximated as $\langle \mathbf{F}\rangle(\omega) = \langle \mathbf{F}_{\text{res}}\rangle(\omega)$.

5.2.4 Expressions for Simple Cases

If we use the basis that diagonalizes only the nonperturbative part in the Hamiltonian that does not include the interaction between the induced polarization and the Maxwell field, the off-diagonal components of the light–matter interaction ($A_{\lambda 0;0\lambda'}$ ($\lambda \neq \lambda'$) in (5.31)) are generally finite. However, depending on the level separations, these components can be neglected as a good approximation. In such a case, $F_\lambda(\omega)$ is described as

$$X_\lambda(\omega) = \frac{X_\lambda^{(\text{b})}(\omega)}{\bar{E}_\lambda - \hbar\omega - i\gamma}$$

$$= \frac{X_\lambda^{(\text{b})}(\omega)}{(E_\lambda + \Delta_\lambda - \hbar\omega) - i(\Gamma_\lambda + \gamma)},$$

$$(5.48)$$

where $\bar{E}_\lambda = E_\lambda + \Delta_\lambda - i\Gamma_\lambda$ is the eigenenergy of the λth light–matter coupled state, $\Delta_\lambda = \text{Re}[A_{\lambda 0;0\lambda}]$ is the radiative shift from the bare eigenenergy E_λ, and $\Gamma_\lambda = -\text{Im}[A_{\lambda 0;0\lambda}]$ is the radiative width (half width at half maximum) of the light–matter coupled state. Hereafter, we derive the analytical expressions of the radiation force under this approximation.

Generally, the incident electric field $\mathbf{E}^{\text{b}}(\mathbf{r},\omega)$ and the dyadic Green function $\mathbf{G}^{\text{b}}(\mathbf{r},\mathbf{r}',\omega)$ in the presence of the background dielectric can be expanded as

$$\mathbf{E}^{\text{b}}(\mathbf{r},\omega) = \sum_\zeta a_\zeta(\omega)\tilde{\mathbf{E}}_\zeta^{\text{b}}(\mathbf{r},\omega_\zeta),$$

$$(5.49)$$

$$a_\zeta(\omega) = \frac{1}{|E_{\text{nom}}^{\text{b}}|^2}\int d\mathbf{r}\tilde{\mathbf{E}}_\zeta^{\text{b}}(\mathbf{r},\omega_\zeta)^* \cdot \mathbf{E}^{\text{b}}(\mathbf{r},\omega),$$

$$(5.50)$$

$$\mathbf{G}^{\text{b}}(\mathbf{r},\mathbf{r}',\omega) = \sum_\zeta \frac{4\pi q^2}{|E_{\text{nom}}^{\text{b}}|^2 \epsilon_{\text{b}}(\mathbf{r})}\frac{\tilde{\mathbf{E}}_\zeta^{\text{b}}(\mathbf{r},\omega_\zeta)^* \tilde{\mathbf{E}}_\zeta^{\text{b}}(\mathbf{r}',\omega_\zeta)}{q_\zeta^2 - q^2}$$

$$(5.51)$$

by the basis set $\{\tilde{\mathbf{E}}_\zeta^b(\mathbf{r}, \omega_\zeta)\}$ satisfying the completeness

$$\int d\mathbf{r}\tilde{\mathbf{E}}_\zeta(\mathbf{r}, \omega_\zeta)^* \cdot \tilde{\mathbf{E}}_{\zeta'}(\mathbf{r}, \omega_{\zeta'}) = |E_{\text{nom}}^b|^2 \delta_{\zeta\zeta'}, \tag{5.52}$$

$$\sum_\zeta \tilde{\mathbf{E}}_\zeta(\mathbf{r}, \omega_\zeta)^* \tilde{\mathbf{E}}_\zeta(\mathbf{r}', \omega_\zeta) = |E_{\text{nom}}^b|^2 \mathbf{I}\delta(\mathbf{r} - \mathbf{r}'), \tag{5.53}$$

where $|E_{\text{nom}}^b|^2$ is the normalization constant, $\epsilon_b(\mathbf{r}) = 1+4\pi\chi_b\Theta(\mathbf{r})$ ($\epsilon_b = 1+4\pi\chi_b$), $\Theta(\mathbf{r})$ satisfies (5.23), and $q_\zeta = \omega_\zeta/c$ is the wave number of the ζth radiation mode (Mie mode) [49, 50]. If we choose the plane wave $\tilde{\mathbf{E}}_\zeta^b(\mathbf{r}, \omega_\zeta) = \exp[i\mathbf{k} \cdot \mathbf{r}]\boldsymbol{\xi}$ as an expansion basis ($\zeta = (\mathbf{k}, \boldsymbol{\xi})$, $\mathbf{k}$: wave vector, $\boldsymbol{\xi}$: unit polarization vector), (5.49)–(5.51) become

$$\mathbf{E}^b(\mathbf{r}, \omega) = \sum_{\mathbf{k},\boldsymbol{\xi}} \mathcal{E}^b(\mathbf{k}, \boldsymbol{\xi}, \omega) \exp[i\mathbf{k} \cdot \mathbf{r}]\boldsymbol{\xi}, \tag{5.54}$$

$$a_\zeta(\omega) = \mathcal{E}^b(\mathbf{k}, \boldsymbol{\xi}, \omega) = \frac{1}{(2\pi)^3} \int d\mathbf{r} \exp[i\mathbf{k} \cdot \mathbf{r}]^* \boldsymbol{\xi} \cdot \mathbf{E}^b(\mathbf{r}, \omega), \tag{5.55}$$

$$\mathbf{G}^b(\mathbf{r}, \mathbf{r}', \omega) = \sum_{\mathbf{k},\boldsymbol{\xi}} \frac{\omega^2}{2\pi^2\epsilon_b(\mathbf{r})} \frac{\exp[i\mathbf{k} \cdot \mathbf{r}]^*\boldsymbol{\xi} \exp[i\mathbf{k} \cdot \mathbf{r}']\boldsymbol{\xi}}{\omega_{\mathbf{k}\boldsymbol{\xi}}^2 - \omega^2}, \tag{5.56}$$

where the normalization constant is $|E_{\text{nom}}^b|^2 = (2\pi)^3$. Thus, $\nabla\mathbf{E}^b(\mathbf{r}, \omega)$, $\nabla\mathbf{G}^b(\mathbf{r}, \mathbf{r}', \omega)$, and $X_\lambda^{(b)}(\omega)$ are expressed as

$$\nabla\mathbf{E}^b(\mathbf{r}, \omega) = \sum_{\mathbf{k}\boldsymbol{\xi}} i\mathbf{k}\mathcal{E}^b(\mathbf{k}, \boldsymbol{\xi}, \omega) \exp[i\mathbf{k} \cdot \mathbf{r}]\boldsymbol{\xi}, \tag{5.57}$$

$$\nabla\mathbf{G}^b(\mathbf{r}, \mathbf{r}', \omega) = \sum_{\mathbf{k},\boldsymbol{\xi}} \frac{-i\mathbf{k}^*\omega^2}{2\pi^2\epsilon_b(\mathbf{r})} \frac{\exp[-i\mathbf{k}^* \cdot \mathbf{r}]\boldsymbol{\xi} \exp[i\mathbf{k} \cdot \mathbf{r}']\boldsymbol{\xi}}{\omega_{\mathbf{k}\boldsymbol{\xi}}^2 - \omega^2}, \tag{5.58}$$

$$X_\lambda^{(b)}(\omega) = \sum_{\mathbf{k},\boldsymbol{\xi}} \mathcal{F}_\lambda^{(b)}(\mathbf{k}, \boldsymbol{\xi}, \omega), \tag{5.59}$$

$$\mathcal{X}_\lambda^{(b)}(\mathbf{k}, \boldsymbol{\xi}, \omega) = \int_V d\mathbf{r}\boldsymbol{\rho}_{\lambda 0}(\mathbf{r}) \cdot \boldsymbol{\xi}\mathcal{E}^b(\mathbf{k}, \boldsymbol{\xi}, \omega) \exp[i\mathbf{k} \cdot \mathbf{r}]. \tag{5.60}$$

Substituting (5.48) and (5.57)–(5.60) into (5.47), we finally obtain the analytical expression of the resonant radiation force for the nano object-confining electronic system as

$$\langle\mathbf{F}\rangle(\omega) = \frac{1}{2}\text{Re}\left[\sum_\lambda\sum_{\mathbf{k},\boldsymbol{\xi}}(-i\mathbf{k}^*)X_\lambda(\omega)\mathcal{X}_\lambda^b(\mathbf{k}, \boldsymbol{\xi}, \omega)^* + \sum_\lambda\sum_{\lambda'}\Xi_{\lambda\lambda'}(\omega)\right]$$

$$= \frac{1}{2}\sum_\lambda\sum_{\mathbf{k},\boldsymbol{\xi}}\sum_{\mathbf{k}',\boldsymbol{\xi}'}\left[\frac{[\mathbf{k}^{(R)}(\Gamma_\lambda + \gamma) - \mathbf{k}^{(I)}(E_\lambda + \Delta_\lambda - \hbar\omega)]W_\lambda^{(R)}(\zeta, \zeta', \omega)}{(E_\lambda + \Delta_\lambda - \hbar\omega)^2 + (\Gamma_\lambda + \gamma)^2}\right.$$

$$\left.+ \frac{[\mathbf{k}^{(R)}(E_\lambda + \Delta_\lambda - \hbar\omega) + \mathbf{k}^{(I)}(\Gamma_\lambda + \gamma)]W_\lambda^{(I)}(\zeta, \zeta', \omega)}{(E_\lambda + \Delta_\lambda - \hbar\omega)^2 + (\Gamma_\lambda + \gamma)^2}\right]$$

$$+ \frac{1}{2}\text{Re}\left[\sum_\lambda\sum_{\lambda'}\Xi_{\lambda\lambda'}(\omega)\right], \tag{5.61}$$

where $\mathbf{k}^{(R)}$ and $\mathbf{k}^{(I)}$ are the real and imaginary parts of $\mathbf{k}$, respectively, and

$$\mathcal{W}_\lambda^{(R)}(\zeta, \zeta', \omega) + i\mathcal{W}_\lambda^{(I)}(\zeta, \zeta', \omega) \equiv \mathcal{X}_\lambda^b(\mathbf{k}, \boldsymbol{\xi}, \omega)^* \mathcal{X}_\lambda^b(\mathbf{k}', \boldsymbol{\xi}', \omega). \quad (5.62)$$

In the above expression, the explicit form of $\boldsymbol{\Xi}_{\lambda\lambda'}(\omega)$, which is in the third term of the right-hand side and provides a small contribution, is very complicated, and we do not present it here. The physical meaning of this additional term is explained later.

By applying (5.61) to simple models, we can understand the general properties of resonant radiation force on nano objects, as demonstrated in the following. We consider the radiation force on a nano object whose center-of-mass is located at $\mathbf{r} = \mathbf{0}$ in the steady state when it is irradiated by (1) a plane wave propagating with a real wave vector $\mathbf{k}_0$ in vacuum ($|\mathbf{k}_0| = q = \omega/c$, $\mathbf{k}^{(R)} = \mathbf{k}_0$, $\mathbf{k}^{(I)} = \mathbf{0}$) and unit polarization vector $\boldsymbol{\xi}$ satisfying the relation of $\boldsymbol{\xi} \perp \mathbf{k}$ (transverse wave) as $\mathbf{E}_{pro}^b(\mathbf{r}, \omega) = \mathbf{E}_+^{pl}(\mathbf{r}, \omega)$, and (2) a standing wave field $\mathbf{E}_{sta}^b(\mathbf{r}, \omega) = \mathbf{E}_+^{pl}(\mathbf{r}, \omega) + \mathbf{E}_-^{pl}(\mathbf{r}, \omega) = 2E^{(0)} \cos[\mathbf{k}_0 \cdot (\mathbf{r} - \mathbf{r}_1)]\boldsymbol{\xi}$, where $\mathbf{E}_\pm^{pl}(\mathbf{r}, \omega) = E^{(0)} \exp[\pm i\mathbf{k}_0 \cdot (\mathbf{r} - \mathbf{r}_1)]\boldsymbol{\xi}$, $\mathbf{r}_1$ is the antinode position of the standing wave, and the scattering from background dielectric is assumed to be negligible.

For the plane wave $\mathbf{E}_{pro}^b(\mathbf{r}, \omega)$, as shown in Fig. 5.2(a), we can easily derive the expression as

$$\langle \mathbf{F} \rangle(\omega) = \sum_\lambda \mathbf{k}_0 \frac{(\Gamma_\lambda + \gamma)(|\mathcal{F}_\lambda^{(b)pro}(\mathbf{k}_0, \boldsymbol{\xi}, \omega)|^2/2)}{(E_\lambda + \Delta_\lambda - \hbar\omega)^2 + (\Gamma_\lambda + \gamma)^2}$$
$$+ \frac{1}{2}\text{Re}\left[\sum_\lambda \sum_{\lambda'} \boldsymbol{\Xi}_{\lambda\lambda'}^{(pl)}(\omega)\right], \quad (5.63)$$

$$\mathcal{F}_\lambda^{(b)pro}(\mathbf{k}_0, \boldsymbol{\xi}, \omega) = \int_V d\mathbf{r}[\rho_{\lambda 0}(\mathbf{r})]_{\boldsymbol{\xi}} E^{(0)} \exp[i\mathbf{k}_0 \cdot \mathbf{r}]. \quad (5.64)$$

For the standing wave field $\mathbf{E}_{sta}^b(\mathbf{r}, \omega)$, as shown in Fig. 5.2(b), we can derive the expression as

$$\langle \mathbf{F} \rangle(\omega) = \sum_\lambda \frac{1}{4} \frac{(E_\lambda + \Delta_\lambda - \hbar\omega)\nabla|\mathbf{E}_{sta}^b(\mathbf{r}, \omega)|^2_{\mathbf{r}=0}(|\mathcal{F}_\lambda^{(b)pro}(\mathbf{k}_0, \boldsymbol{\xi}, \omega)|/E^{(0)})^2}{(E_\lambda + \Delta_\lambda - \hbar\omega)^2 + (\Gamma_\lambda + \gamma)^2}$$
$$+ \frac{1}{2}\text{Re}\left[\sum_\lambda \sum_{\lambda'} \boldsymbol{\Xi}_{\lambda\lambda'}^{(st)}(\omega)\right]. \quad (5.65)$$

The above expressions, namely, (5.63), (5.64) and (5.65), reproduce all the properties explained by the conventional expressions of radiation force from the classical regime to the atomic regime, namely, (5.4), (5.5) and (5.6) for the former regime, and (5.8) and (5.9) for the latter regime. Furthermore, these expressions include generalizations corresponding to the features specific for nanoscale systems. The classical expression of dissipative force (5.4) includes scattering force and absorbing force. These kinds of forces are expressed by (5.63), namely the portions proportional to Γ_λ and to γ in the numerator of the first term mean the scattering force

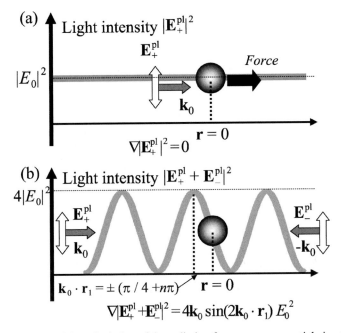

Fig. 5.2. a Geometry of the calculation of the radiation force on a nanoparticle irradiated by a propagating plane wave. **b** Geometry of the calculation of the radiation force on a nanoparticle in a standing wave field

and the absorbing one, respectively. These terms are also proportional to the intensity of the incident field. The effect of the anisotropic factor $(1 - \overline{\cos\theta})$ in (5.4) can be reproduced in a more general form by considering the second term including $\mathbf{\Xi}^{(pl)}_{\lambda\lambda'}(\omega)$. In the classical expression for a point dipole (5.6), the gradient force appears, which is proportional to the gradient of the incident intensity. If we neglect the anisotropic scattering considering the object as a point dipole, the present expressions (5.63), (5.64) and (5.65) are consistent with (5.6). In contrast with (5.6), the effect of the spatial extension of the electronic system and its resonant structure are explicitly considered through the transition dipole moment proportional to the excitonic wavefunction, $\boldsymbol{\rho}_{\lambda 0}(\mathbf{r})$, and the energy denominator, respectively, in the present expressions. Also, we can see that the present expressions are generalizations of those for an atom calculated using the Heisenberg equation of atomic translational motion under long-wavelength approximation, namely, the first term on the right-hand side of (5.63) corresponds to the dissipative force, and the first term on the right-hand side of (5.65) corresponds to the reactive force (gradient force) on an atom in (5.8) and (5.9) at low intensity (under the condition that $|\Omega_1|^2$ in the denominator can be neglected). The present expressions explicitly include the radiative correction that becomes remarkable for the comparatively large nano objects. Namely, (5.63) explains that the frequency spectrum of dissipative force has a peak value $\propto |(\mathcal{F}^{(b)pro}_\lambda(\mathbf{k}_0, \boldsymbol{\xi}, \omega)|/(\Gamma_\lambda + \gamma)$ at an energy of $E_\lambda + \Delta_\lambda$ and (5.65) shows that the sign of reactive force changes at this energy. Furthermore, remarkable points are that the absolute square of Rabi frequency $|\Omega_1|^2$ for a point dipole is replaced by a size dependent quantity $|\mathcal{F}_\lambda^{(b)pro}(\mathbf{k}, \boldsymbol{\xi}, \omega)/\hbar|^2$, which reflects not only the amplitude of the incident electric field but also the microscopic spatial structure of the confined electronic system.

The additional term $\boldsymbol{\Xi}_{\lambda\lambda'}(\omega)$ appears from the anisotropy of the scattering. In contrast with the term $(1 - \overline{\cos\theta})$ in (5.4), this additional term includes the spatial structures of the electronic system, namely, it arises due to the contributions of several levels with different parities. In the case of a spherical object, for example, $\boldsymbol{\Xi}_{\lambda\lambda'}(\omega) \propto a^6$, where a is the sphere radius. Therefore, for $a \to 0$, $\boldsymbol{\Xi}_{\lambda\lambda'}(\omega)$ can be neglected because $|\mathcal{F}_\lambda^{(b)pro}(\mathbf{k}, \boldsymbol{\xi}, \omega)/\hbar|^2 \propto a^3$. However, for a particle with a radius of several tens of nanometers, the magnitude of $\boldsymbol{\Xi}_{\lambda\lambda'}(\omega)$ remarkably increases because of an enhancement of the induced polarizations with an asymmetric spatial pattern from the nonlocality of the response. Particularly, for such a particle, this force greatly contributes as negative force under certain conditions even if the incident light is a propagating plane wave without intensity gradient. We can understand that the origin of this force is similar to a gradient force arising from the asymmetric spatial distribution of internal field. (See [51] for detailed discussions on this topic.)

5.3 Radiation Force on a Single Nanoparticle Confining Excitons

In the previous section, we derived general analytical expressions of the radiation force on nano objects. Those expressions are very useful for discussing the fundamental features of radiation force in the nanoscale regime by comparing them with the known expressions for the classical and atomic regimes. Further, these expressions can be extended to the system including many nano objects, which provides the essential information on the radiation-induced interparticle force. (For detailed studies on this topic, see [52] and [53].)

In this section, we numerically evaluate the radiation force on a spherical semiconductor particle in order to reveal the features of resonant radiation force peculiar to the nanoscale regime and the feasibility of nanoscale optical manipulation [54–56]. As a model of the matter system, we consider a spherical semiconductor particle with confining of the excitonic center-of-mass motion. Although the expressions derived in the previous section are very useful for discussing the fundamental aspects of radiation force, it is better to use a complementary approach in order to secure the accuracy of numerical calculation for the size regime including larger particles where not a few quantum levels contribute to the optical response. Therefore, we show a different method of solving microscopic Maxwell's equations with nonlocal susceptibility in the linear response regime, namely, the additional boundary condition (ABC) method that is discussed in [57–60]. Although this method is based on the phenomenological treatment of the boundary conditions of the induced polarization, it is known that it provides physically the same results as those provided by the microscopic calculations for a particular simple model of confined excitons and is useful for the numerical calculation of the resonant optical response of the

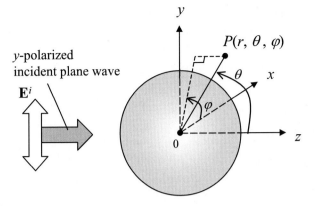

Fig. 5.3. Plane wave incidence on a dielectric sphere in the spherical coordinate system

semiconductor materials that confine excitonic center-of-mass motion [59, 60]. This approach allows a rigorous calculation even for a larger system, though it is not easy to identify the respective contributions from the individual quantum states of a large system.

As shown in Fig. 5.3, we assume that a spherical particle is irradiated by a y-polarized plane wave propagating in the z-direction, where the particle is assumed to be in a medium with unit dielectric constant, such as vacuum or superfluid He^4. Assuming that the center of the sphere is located at the origin $(x, y, z) = (0, 0, 0)$, we expand the fields using spherical surface harmonics and solve the boundary conditions including ABCs to derive the response field. (The details of the expansion of the fields using the spherical surface harmonics are shown in [28].) This method is based on the wave number (k) dependent dielectric function (with spatial dispersion) of the light-exciton coupled oscillator (exciton-polariton) in the system with a translational symmetry as

$$\epsilon(\omega, k) = \epsilon_b + \frac{\epsilon_b \Delta_{LT}}{\hbar\omega_T - \hbar\omega + Dk^2 - i\gamma},$$

$$D = \frac{\hbar^2}{2M_{ex}}, \tag{5.66}$$

where ϵ_b is the background dielectric constant, Δ_{LT} is the LT splitting energy, $\hbar\omega_T$ is the transverse energy of bulk exciton, ω is the frequency of incident light, and M_{ex} is the translational mass of exciton. γ is the phenomenologically introduced nonradiative damping parameter. By solving the dispersion equation $(\omega^2/c^2)\epsilon(\omega, k) = k^2$ using the dielectric function (5.66), we can determine the wave number k.

(i) For transverse waves:

$$k^4 + \left[\frac{2M_{ex}}{\hbar^2}(\hbar\omega_T - \hbar\omega - i\gamma) - \epsilon_b\frac{\omega^2}{c^2}\right]k^2$$

$$- \epsilon_b\frac{2M_{ex}\omega^2}{\hbar^2c^2}(\hbar\omega_T - \hbar\omega - i\gamma + \Delta_{LT}) = 0. \tag{5.67}$$

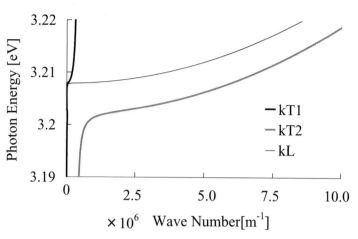

Fig. 5.4. Dispersion curves of bulk exciton–polariton

The physically meaningful solutions of (5.67) are restricted to those that have a positive imaginary part; otherwise, the coordinate dependent part in a plane wave, namely,

$$e^{ikr} = e^{iRe(kr)}e^{-Im(kr)}$$

diverges infinitely as r increases. Thus, two transverse wave numbers k_{T1} and k_{T2} are obtained.

(ii) For longitudinal wave:

The roots of $\epsilon(\omega, k) = 0$ provide the solutions for the longitudinal waves as

$$k = \pm\sqrt{-\frac{2M_{ex}}{\hbar^2}(\hbar\omega_T - \hbar\omega - i\gamma + \Delta_{LT})}, \qquad (5.68)$$

where we determine k_L as solutions that have a positive imaginary part.

The relation between the incident frequency ω and the wave numbers (k_{T1}, k_{T2}, k_L) is shown in Fig. 5.4, where we use the parameters of CuCl Z_3 exciton as $\epsilon_b = 5.59$, $\Delta_{LT} = 5.7$ [meV], $\omega_T = 3.2022$ [eV], and $M_{ex} = 2.3m_0$ (m_0 is the mass of a free electron). γ is assumed to be 20 [μeV]. From Fig. 5.4, we can see that k becomes a multivalued function of ω. The branch starting from $k = 0$, $\omega = 0$ is called "lower branch" (thick grey line) and another transverse branch "upper branch" (thick black line). It can be understood that plural modes with different wave numbers are propagating in the medium above the longitudinal exciton energy. To determine the unique relation between the amplitudes of the inner and the outer fields, the usual Maxwell boundary conditions

$$\left(\mathbf{E}^{(i)} + \mathbf{E}^{(s)}\right)_t = \left(\mathbf{E}^{(T1)} + \mathbf{E}^{(T2)} + \mathbf{E}^{(L)}\right)_t, \qquad (5.69)$$

$$\left(\mathbf{H}^{(i)} + \mathbf{H}^{(s)}\right)_t = \left(\mathbf{H}^{(T1)} + \mathbf{H}^{(T2)}\right)_t \qquad (5.70)$$

are not sufficient due to the existence of these plural propagating waves with k_{T1}, k_{T2}, k_L, where superscript (i) indicates the incident field and (s) indicates the scattered field; (T1), (T2) and (L) are the internal fields with wave numbers k_{T1}, k_{T2} and k_L, respectively; and subscript t indicates the tangential components at the surface. In this case, for obtaining the response fields, we need to use the ABCs. The Pekar-type ABCs [57], for example, require the excitonic polarizations to vanish at the surface ($\mathbf{P} = 0$) as

$$\left(\epsilon(\omega, k_{T1}) - \epsilon_b\right)\mathbf{E}_t^{(T1)} + \left(\epsilon(\omega, k_{T2}) - \epsilon_b\right)\mathbf{E}_t^{(T2)} - \epsilon_b\mathbf{E}_t^{(L)} = 0, \qquad (5.71)$$

$$\left(\epsilon(\omega, k_{T1}) - \epsilon_b\right)\mathbf{E}_n^{(T1)} + \left(\epsilon(\omega, k_{T2}) - \epsilon_b\right)\mathbf{E}_n^{(T2)} - \epsilon_b\mathbf{E}_n^{(L)} = 0, \qquad (5.72)$$

where subscript n indicates the normal components at the surface. This treatment is equivalent to considering the nonescape boundary condition for the center-of-mass wave functions of the exciton [58, 60]. By solving the above Maxwell boundary conditions and ABCs, the microscopic response field can be obtained. If we substitute this solution in (5.16), the radiation force from the microscopic calculation is obtained.

5.3.1 Size Dependence

In the following calculations, we consider the two types of incident fields, i.e., the propagating plane wave $\mathbf{E}_+^i(z)$ and the standing wave $\mathbf{E}_+^i(z) + \mathbf{E}_-^i(z)$, where $\mathbf{E}_\pm^i(z) = E_0(0, \exp[\pm ik_0(z - z_1)], 0)\exp[-i\omega t]$ and k_0 is the wavenumber in a vacuum. In the former case, both the scattering and the absorbing forces are induced, while the gradient force (dipole force) is induced in the latter case. For the numerical demonstration, the incident intensity is assumed to be 50 [μW/100 μm^2] ($=$ 50 [W/cm^2], and $|E_0|^2 = 12\pi \times 10^7$ [V^2/m^2]), which is within the linear response regime [61].

5.3.1.1 For a Propagating Plane Wave

In Figs. 5.5–5.7, we investigate the size dependence of the frequency spectra of the radiation force on a dielectric sphere in the presence of (Resonance) and in the absence of (Non-res.) the resonance term in the wave number-dependent dielectric function (5.66) when the sphere is irradiated by a propagating plane wave in the geometry as shown in Fig. 5.3. Although the peaks due to the Mie resonance arising from the background dielectric are dominant in the case of a sphere of radius larger than 100 nm, these peaks move to the higher energy region, and the exerted force due to the Mie scattering becomes very small in the frequency region including the available laser frequency. This decrease in the force is one of the reasons for the difficulty in manipulating nano objects. In contrast to the Mie resonance, the peaks due to the excitonic resonance maintain their magnitude and position even if the particle size decreases.

Next, we plot the radiation force normalized by the cross-section of the sphere (πa^2) as a function of the laser frequency for several radii extending the range of the frequency from 0 to 10 eV in order to investigate the properties of the force in detail

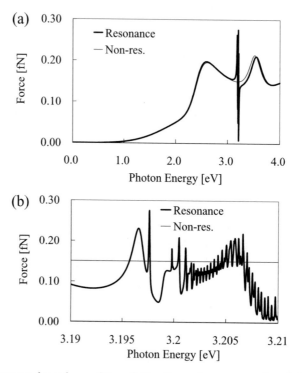

Fig. 5.5. a Frequency dependence of the radiation force when a target sphere is irradiated by a propagating plane wave: Radius $= 100$ [nm]. The result in the absence of the resonance term is also shown. **b** Enlarged view near the resonance energy

(Fig. 5.8(a)). Peaks due to the Mie scattering move to the higher energy region while maintaining their magnitudes, which suggests that each peak value of the force due to Mie resonance is proportional to the surface area of the sphere $(4\pi a^2)$ reflecting the spatial structure of a Mie mode (whispering gallery mode). On the other hand, we convert the force into acceleration (force normalized by the mass of the sphere; mass density of CuCl is $4.18534 \, g/cm^3$) and plot it as a function of frequency for several radii of sphere to investigate the property of these peaks in more detail as shown in Fig. 5.8(b) (normalizing the force by the mass means the evaluation of the force per unit volume). The acceleration peaks originating from the excitonic resonance become larger as the radius decreases when it is less than 100 nm and saturate at last.

In Fig. 5.9, we compare the size dependence of the maximum value of the acceleration induced by the radiation force in 0–4 eV with the gravitational acceleration. Under the excitonic resonance condition, the acceleration curve of a particle of near 10-nm radius becomes parallel to that of the gravitational acceleration. This indicates that the size dependence of the force changes from that proportional to the surface area $(\propto a^2)$ to that proportional to the volume $(\propto a^3)$. This tendency can be understood from the spatial distribution of the electric field inside and outside of the sphere as shown in Fig. 5.20 of Sect. 5.3.2. On the other hand, because the peaks of

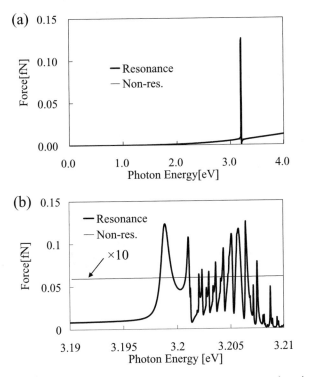

Fig. 5.6. a Frequency dependence of the radiation force when a target sphere is irradiated by a propagating plane wave: Radius = 50 [nm]. The result in the absence of the resonance term is also shown. **b** Enlarged view near the resonance energy

the force due to the Mie resonance are absent in the 0–4 eV region when the radius is smaller than 50 nm, the force due to the Rayleigh scattering is dominant under the nonresonant condition. In this case, the force has the maximum value at 4 eV and decreases proportional to a^6 following

$$F_{scat} = \frac{128\pi^5 a^6}{3\lambda^4}\left(\frac{m^2-1}{m^2+2}\right)I\frac{n_b}{c} \propto a^6, \qquad (5.73)$$

where n_a and n_b are the refractive indices inside and outside the sphere, respectively, $(m = n_a/n_b)$; λ is the wavelength of the incident light; and I is the intensity of the incident light [14]. Thus, fixing the wavelength λ under nonresonant condition, the acceleration of a particle of radius smaller than 50 nm decreases proportional to a^3 as shown in Fig. 5.9.

5.3.1.2 For a Standing Wave Field

Here, we calculate the force exerted on a spherical particle in the standing wave field consisting of two plane waves propagating in mutually opposite directions along the

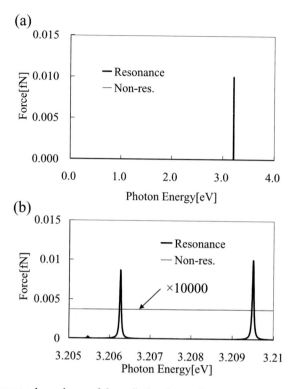

Fig. 5.7. a Frequency dependence of the radiation force when a target sphere is irradiated by a propagating plane wave: Radius = 10 [nm]. The result in the absence of the resonance term is also shown. **b** Enlarged view near the resonance energy

z-axis, where the gradient force proportional to the gradient of the field intensity is dominant (first term in (5.65)). We assume that the particle's center-of-mass is located at $z = 0$ and the antinode position of the standing wave is $z_1 = -\lambda/8$ where the gradient of the incident field intensity has the maximum value (see Fig. 5.10). In the same way as for a single-plane wave irradiation, we investigate the behaviors of the force on the particle of radius from 10 nm to 1000 nm.

The characteristic behavior of the gradient force appears more clearly in the case of a smaller radius because the separation between the quantized levels is larger. In Fig. 5.11, we can see that the sign of the force changes at the center frequencies of the Lorentzian curve of the force on a particle irradiated by a plane wave, which corresponds to the quantized excitonic energies including the radiative shift. The induced force pushes the particle toward the strong intensity region (antinode position) when the incident photon energy is less than the resonance energy, but toward the weak intensity region (node position) when it exceeds the resonance energy. On the other hand, in the case of a larger radius, complicated spectral peak structures appear between $\hbar\omega_T$ and $\hbar\omega_L$ due to the contribution of the background dielectric and interference between plural resonance levels as shown in Figs. 5.12–5.13.

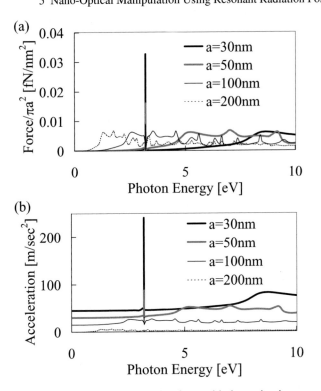

Fig. 5.8. a Frequency dependence of the radiation force with the excitonic resonance term for several radii (normalized by the cross-section of the sphere πa^2). **b** Frequency dependence of the acceleration with the excitonic resonance term for several radii (force normalized by the mass of the sphere)

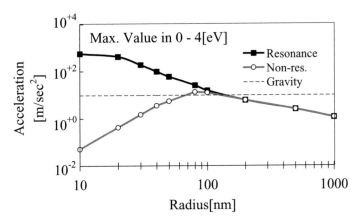

Fig. 5.9. Size dependence of the maximum value of the acceleration induced by radiation force on a spherical particle irradiated by a plane wave in the energy range of 0–4 eV. The gravitational acceleration is also shown

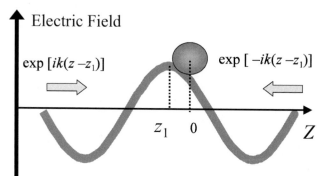

Fig. 5.10. Geometry of a spherical particle located in a standing wave field consisting of two plane waves propagating in mutually opposite directions

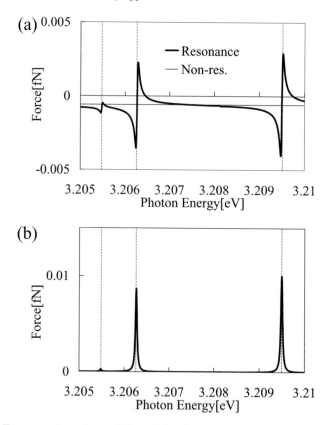

Fig. 5.11. a Frequency dependence of the radiation force when a target sphere is in the standing wave field: Radius = 10 [nm]. The result in the absence of the resonance term is also shown (enlarged view near the resonance energy). **b** Frequency dependence of the radiation force when a target sphere is irradiated by a plane wave field: Radius = 10 [nm]. Vertical broken lines indicate the positions of the excitonic resonance energies including the radiative shift

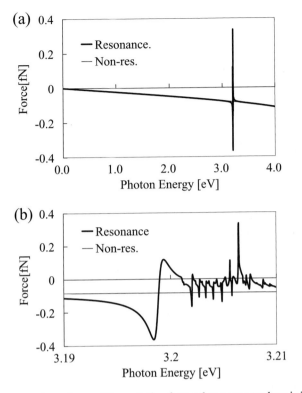

Fig. 5.12. a Frequency dependence of the radiation force when a target sphere is in the standing wave field: Radius $= 50$ [nm]. The result in the absence of the resonance term is also shown. **b** Enlarged view near the resonance energy

In Fig. 5.14, we compare the size dependence of the maximum value of the acceleration in the 0–4 eV region with the gravitational one. In this case, again, the radiation force under the excitonic resonance induces much greater acceleration than that in the absence of the resonance term (Non-res.) and the gravitational acceleration when the particle radius is smaller than 100 nm, and the exciton-induced force is proportional to the volume of the sphere in the small-radius region. This case is different from that of plane wave irradiation in that the force in the absence of the resonance term is also proportional to the volume of the sphere in the Rayleigh scattering regime (where the radius of the particle is much smaller than the light wavelength). This can be explained by the following equation of the gradient force on a Rayleigh particle

$$F_{\text{grad}} = \frac{n_{\text{b}}}{2}\alpha \nabla E_i^2,$$

$$\alpha = \frac{n_{\text{b}}^2 a^3}{2}\left(\frac{m^2-1}{m^2+2}\right) \propto a^3, \tag{5.74}$$

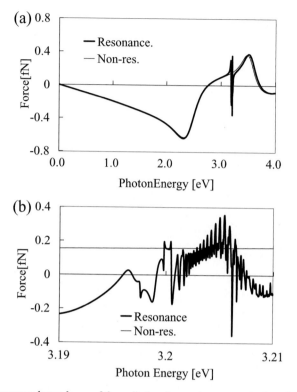

Fig. 5.13. a Frequency dependence of the radiation force when a target sphere is in the standing wave field: Radius = 100 [nm]. The result in the absence of the resonance term is also shown. **b** Enlarged view near the resonance energy

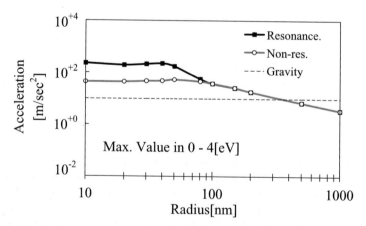

Fig. 5.14. Size dependence of the maximum value of the acceleration induced by radiation force on a spherical particle in the standing wave field in the energy range of 0–4 eV. The gravitational acceleration is also shown

which is proportional to a^3 ($m = n_a/n_b$; n_a and n_b are the refractive indices inside and outside of the sphere, respectively; λ is the wavelength of the incident light; and E_i is the amplitude of the incident light) [14]. Therefore, fixing λ to the wavelength corresponding to 4 eV, the acceleration of a sphere of radius smaller than 50 nm remains constant in magnitude even when the radius decreases.

5.3.2 Several Types of Forces

In the previous section, we discussed the size dependence of the resonant radiation force, where we mainly observed the effects of the dissipative force and the gradient force when the target sphere is irradiated by a propagating wave and a standing wave, respectively. In this section, we focus on how the scattering force and the absorbing force contribute to the radiation force. Generally, the dissipative force includes the scattering and absorbing components. In the resonant radiation force, the ratio of these different types of forces shows complicated frequency dependence specific to confined systems. Before discussing this issue, we consider how the scattering and absorption cross-sections contribute to the optical response, and how these processes appear in the radiation force based on the classical theory.

5.3.2.1 Extinction, Scattering and Absorption Cross-sections

Here, we show the extinction, scattering and absorption cross-sections introduced by the Mie theory [28, 48].

Assuming the Poynting vectors of the incident, scattered, interfered and total external fields as $\mathbf{S}^{(i)}$, $\mathbf{S}^{(s)}$, $\mathbf{S}^{(ext)}$ and $\mathbf{S}^{(1)}$, we can write these quantities as

$$\mathbf{S}^{(i)} = \frac{1}{2}\text{Re}\left\{\mathbf{E}^{(i)} \times \mathbf{H}^{(i)*}\right\}, \tag{5.75}$$

$$\mathbf{S}^{(s)} = \frac{1}{2}\text{Re}\left\{\mathbf{E}^{(s)} \times \mathbf{H}^{(s)*}\right\}, \tag{5.76}$$

$$\mathbf{S}^{(ext)} = \frac{1}{2}\text{Re}\left\{\mathbf{E}^{(i)} \times \mathbf{H}^{(s)*} + \mathbf{E}^{(s)} \times \mathbf{H}^{(i)*}\right\}, \tag{5.77}$$

$$\mathbf{S}^{(1)} = \frac{1}{2}\text{Re}\left\{\mathbf{E}^{(1)} \times \mathbf{H}^{(1)*}\right\} = \mathbf{S}^{(i)} + \mathbf{S}^{(s)} + \mathbf{S}^{(ext)}, \tag{5.78}$$

respectively.

As illustrated in Fig. 5.15, the difference between the light energy received by the detector in the presence of a particle and that in the absence of the particle is caused by absorption, which is the conversion of the energy of electromagnetic field into another form such as heat in the matter and coherent light scattering. As a whole, we call these phenomena "extinction". This extinction generally depends on the composition of an object, its shape, spatial configuration, environment, the polarization and wavelength of the incident light.

Considering a particle in nonabsorbing medium and imaginary spherical surface A around it as shown in Fig. 5.15, the total electromagnetic energy that crosses surface A of this sphere can be written as

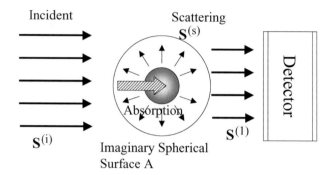

Fig. 5.15. Schematic view of the extinction by a single particle

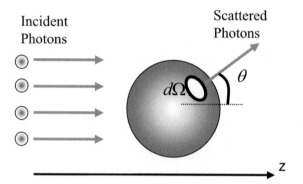

Fig. 5.16. Incident and scattered photons

$$W^{(1)} = -\int_A \mathbf{S}^{(1)} \cdot \hat{\mathbf{e}}_r \, dS. \tag{5.79}$$

Since the medium is nonabsorbing, this implies that $W^{(1)}$ is the ratio of the energy absorbed by the particle when $W^{(1)} > 0$ is satisfied. Rewriting as $W^{(1)} = W^{(abs)}$, each energy can be described as

$$W^{(abs)} = W^{(inc)} - W^{(s)} + W^{(ext)}, \tag{5.80}$$

$$W^{(inc)} = -\int_A \mathbf{S}^{(i)} \cdot \hat{\mathbf{e}}_r \, dS, \tag{5.81}$$

$$W^{(scat)} = \int_A \mathbf{S}^{(s)} \cdot \hat{\mathbf{e}}_r \, dS, \tag{5.82}$$

$$W^{(ext)} = -\int_A \mathbf{S}^{(ext)} \cdot \hat{\mathbf{e}}_r \, dS, \tag{5.83}$$

where $W^{(inc)} = 0$ leads to

$$W^{(ext)} = W^{(abs)} + W^{(scat)}. \tag{5.84}$$

$\sigma_{scat} = W^{(scat)}/I^{(i)}$ ($I^{(i)}$: incident intensity) is called the total "scattering cross-section" that can be expressed as

$$\sigma_{\text{scat}} = W^{(\text{scat})}/I^{(\text{i})} = \lim_{r\to\infty} r^2 \int \frac{|\mathbf{E}^{(\text{s})}(\mathbf{r})|^2}{|\mathbf{E}^{(\text{i})}(\mathbf{r})|^2} d\Omega, \tag{5.85}$$

where the particle is irradiated by a single plane wave light. We know the net ratio of the scattered intensity to the incident intensity by evaluating this quantity.

On the other hand, we can evaluate the "extinction cross-section" by calculating

$$\sigma_{\text{ext}} = W^{(\text{ext})}/I^{(\text{i})} = \lim_{r\to\infty} r^2 \int \frac{|\mathbf{E}^{(\text{ext})}(\mathbf{r})|^2}{|\mathbf{E}^{(\text{i})}(\mathbf{r})|^2} d\Omega, \tag{5.86}$$

which implies the net ratio of the extinction energy to the total energy of the incident light. Finally, we obtain the "absorption cross-section" by calculating the difference between σ_{ext} and σ_{scat} as

$$\sigma_{\text{abs}} = W^{(\text{abs})}/I^{(\text{i})} = \sigma_{\text{ext}} - \sigma_{\text{scat}}, \tag{5.87}$$

which implies the net ratio of the intensity absorbed by the particle to the incident intensity. (Note that we plot σ_{scat}, σ_{ext} and σ_{abs} normalized by πa^2 in the latter numerical evaluations of Figs. 5.17–5.19.)

5.3.2.2 Interpretation of the Scattering and the Absorbing Force Based on the Classical Mechanics

When the incident light is a single propagating plane wave, the scattering force and the absorbing force are dominant in the radiation force because the field is homogeneous and there is no intensity gradient. In this case, the expression of the force is described as

$$\langle F_z \rangle = \frac{1}{8\pi} \left(\sigma_{\text{ext}} - \sigma_{\text{scat}} \overline{\cos\theta} \right) |E^{(\text{i})}|^2 \tag{5.88}$$

by the Maxwell stress tensor method [27, 28] and the classical mechanics. Since the former derivation is complicated, here we show an easy method to derive this expression based on the energy and momentum conservation laws in classical mechanics.

Now, we consider that the incident light is a group of photons with an energy of $\hbar\omega$ per photon and that the number of photons crossing an imaginary spherical surface A surrounding the dielectric particle per unit time is dN/dt. Assuming that the total energy of these photons is divided into the kinetic energy of the matter and the energies of scattering and absorption, the energy and z-component of the momentum are conserved as

$$\frac{dN}{dt}\hbar\omega = \frac{d}{dt}\left(\frac{1}{2}m_s v_s^2\right) + \int \frac{dn'(\Omega)}{dt}\hbar\omega\, d\Omega + \frac{d\mathcal{E}_{\text{abs}}}{dt}, \tag{5.89}$$

$$\frac{dN}{dt}\frac{\hbar\omega}{c} = \frac{d}{dt}(m_s v_s) + \int \frac{dn'(\Omega)}{dt}\frac{\hbar\omega}{c}\cos\theta\, d\Omega, \tag{5.90}$$

where m_s is the mass of the target particle, v_s is the z-component of the velocity of the particle, $n'(\Omega)$ is the number of scattered photons per unit steradian, θ is the scattering angle between the propagation direction of a scattered photon and the z-axis, and $d\mathcal{E}_{\text{abs}}/dt$ is the energy absorbed by the matter per unit time. The average number of scattered photons crossing a small surface element, which subtends a solid angle $d\Omega$ steradian, per unit time is

$$N' = \int n'(\Omega)\,d\Omega, \qquad (5.91)$$

and the average cosine of the scattered angle is

$$\overline{\cos\theta} = \frac{\int n'(\Omega)\cos\theta\,d\Omega}{\int n'(\Omega)\,d\Omega}. \qquad (5.92)$$

Substituting (5.91) and (5.92) into the simultaneous equations (5.89) and (5.90) and solving them leads to

$$\frac{dN}{dt}\hbar\omega = v_{\text{s}}\frac{d}{dt}(m_{\text{s}}v_{\text{s}}) + \frac{dN'}{dt}\hbar\omega + \frac{d\mathcal{E}_{\text{abs}}}{dt}, \qquad (5.93)$$

$$\frac{dN}{dt}\hbar\omega = c\frac{d}{dt}(m_{\text{s}}v_{\text{s}}) + \frac{dN'}{dt}\hbar\omega\overline{\cos\theta}. \qquad (5.94)$$

Since the component of the incident direction of force F_z is

$$F_z = \frac{d}{dt}(m_{\text{s}}v_{\text{s}}), \qquad (5.95)$$

we obtain the expression of the radiation force as

$$F_z = \frac{1}{c - v_{\text{s}}}\frac{d}{dt}\big[N'\hbar\omega\big(1 - \overline{\cos\theta}\big) + \mathcal{E}_{\text{abs}}\big]. \qquad (5.96)$$

Now, we can replace the time derivative of the energy of the scattered photons by the surface integral of the Poynting vector over the imaginary spherical surface A as

$$\frac{dN'}{dt}\hbar\omega = \int_A \mathbf{S}^{(\text{s})}\cdot\mathbf{n}\,dS = \sqrt{\frac{\epsilon}{\mu}}\sigma_{\text{scat}}\big|E^{(\text{i})}\big|^2. \qquad (5.97)$$

Similarly, using

$$\frac{d\mathcal{E}_{\text{abs}}}{dt} = \sqrt{\frac{\epsilon}{\mu}}\sigma_{\text{abs}}\big|E^{(\text{i})}\big|^2, \qquad (5.98)$$

we can rewrite the expression of the force as

$$F_z = \epsilon_1\big\{\big(\sigma_{\text{scat}}\big(1 - \overline{\cos\theta}\big) + \sigma_{\text{abs}}\big)\big\}\big|E^{(\text{i})}\big|^2 = \epsilon_1\big(\sigma_{\text{ext}} - \sigma_{\text{scat}}\overline{\cos\theta}\big)\big|E^{(\text{i})}\big|^2, \qquad (5.99)$$

where we use the relations $v_{\text{s}} \ll c$, $\frac{1}{c} = \sqrt{\epsilon\mu}$, and $\epsilon = \epsilon_1$. Taking the time average of the force, we obtain $\langle F_z\rangle = (1/2)\epsilon_1(\sigma_{\text{ext}} - \sigma_{\text{scat}}\overline{\cos\theta})\langle|E^{(\text{i})}|^2\rangle$, and we can rewrite this as

$$\langle F_z \rangle = \frac{1}{8\pi} \epsilon_1 \{ (\sigma_{\text{scat}}(1 - \overline{\cos\theta}) + \sigma_{\text{abs}}) \} |E^{(i)}|^2$$
$$= \frac{1}{8\pi} (\sigma_{\text{ext}} - \sigma_{\text{scat}}\overline{\cos\theta}) |E^{(i)}|^2 \tag{5.100}$$

in CGS units ($\epsilon_1 = 1$).

When the scattering is isotropic, $\overline{\cos\theta} \cong 0$ and the force is proportional to the extinction cross-section σ_{ext}. Moreover, we can compare the ratio of the contributions of scattering and absorption to the force as

$$\langle F_{\text{scat}} \rangle = \frac{dN_{\text{scat}}}{dt} \frac{\hbar\omega}{c} = \frac{1}{8\pi} \sigma_{\text{scat}} |E^{(i)}|^2, \tag{5.101}$$

$$\langle F_{\text{abs}} \rangle = \frac{1}{c} \frac{d\mathcal{E}_{\text{abs}}}{dt} = \frac{dN_{\text{abs}}}{dt} \frac{\hbar\omega}{c} = \frac{1}{8\pi} \sigma_{\text{abs}} |E^{(i)}|^2. \tag{5.102}$$

Therefore, the ratio between the scattering force and the absorbing force becomes

$$\langle F_{\text{scat}} \rangle : \langle F_{\text{abs}} \rangle = \frac{dN_{\text{scat}}}{dt} : \frac{dN_{\text{abs}}}{dt} = \sigma_{\text{scat}} : \sigma_{\text{abs}}, \tag{5.103}$$

which is the ratio between the scattering and absorption cross-sections. When the scattering is anisotropic (in the case that a large sphere or an irregular-shaped object is irradiated by a plane wave), the contribution of $\overline{\cos\theta}$ cannot be neglected and the character of the force is not explained by such a simple relation.

5.3.2.3 Comparison of Scattering, Absorption, Extinction Cross-Sections and Radiation Force by Numerical Calculation

Previously, we explained that when a particle is irradiated by a propagating plane wave, the main part of the radiation force is divided into two components: one attributable to the coherent scattering and the other to the nonradiative absorption. Here, we methodically investigate their relation by numerical calculation. In the case of a large sphere, the contribution of $\overline{\cos\theta}$ cannot be neglected due to the anisotropic scattering where the force is not proportional to the extinction cross-section σ_{ext} as a sum of scattering cross-section σ_{scat} and absorption cross-section σ_{abs}. For a smaller particle, since the contribution of $\overline{\cos\theta}$ becomes relatively small, the spectrum of the force approaches that of σ_{ext} (Figs. 5.17–5.19).

For a particle with a radius of several tens of nanometers, the radiative widths of some quantized levels becomes large as compared with the nonradiative width due to the strong radiation-exciton coupling [60], where the ratio of nonradiative absorption (σ_{abs}) becomes relatively smaller than the scattering (σ_{scat}) (e.g., in the frequency region just below $\hbar\omega_T$ ($= 3.2022\,\text{eV}$) or the surface mode between $\hbar\omega_T$ and $\hbar\omega_L$ as shown in Fig. 5.17). In the case of the resonant excitation, usually, the absorption cannot be neglected; hence, the heating problem might be serious for optical manipulation. However, if the frequency of the incident light corresponds to the eigenenergies of these wide peaks (indicated by arrows $\langle i \rangle$, $\langle iii \rangle$ and $\langle iv \rangle$ in Fig. 5.17), the force is dominated by elastic light scattering. In such a case, heating is not serious because the radiative decay is much faster than nonradiative decay.

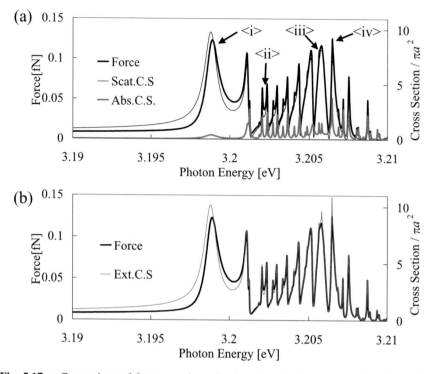

Fig. 5.17. a Comparison of frequency dependencies of scattering cross-section, absorption cross-section and radiation force: Radius = 50 [nm]. **b** Comparison of frequency dependencies of extinction cross-section and radiation force: Radius = 50 [nm]

For a smaller radius, for example, 10 nm, the absorbing force becomes dominant as the radiative width decreases (Fig. 5.18). However, in the case of small damping (1.0 µeV as reported in [62]), as shown in Fig. 5.19, the peak value of the force is enhanced and the ratio of the scattering force to the absorbing one becomes large even in the case of such a small particle while the integrated intensity of the force is the same as that for large damping. Therefore, using the resonant laser whose width is narrower than that of the peak of a quantized level, it would be possible to separate only high-quality particles with small damping by the strong force arising under the conditions of small absorption and less heating.

5.3.2.4 Internal and External Electric Field Distributions

The peak-specific feature of the force shown above is related to the nanoscale spatial structures of the internal field. Here, we show the spatial distributions of the absolute value of the electric field (normalized by that of the incident light) on the xz-plane ($y = 0$) inside and outside of the sphere. We choose the energies of the peaks with the specific feature of the force in its spectra. As shown in Fig. 5.20(a), a sphere is located at the center of each figure and enclosed within a white circle. The incident

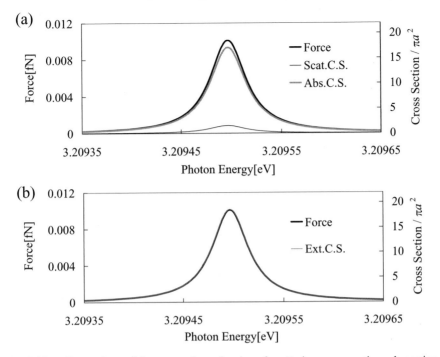

Fig. 5.18. a Comparison of frequency dependencies of scattering cross-section, absorption cross-section and radiation force: Radius = 10 [nm]; γ = 20 [μeV]. **b** Comparison of frequency dependencies of extinction cross-section and radiation force: Radius = 10 [nm]; γ = 20 [μeV]

plane wave light propagates in the z-direction (the light polarization is perpendicular to the paper; y-polarization).

If a resonant light is irradiated, the internal light field with shortened wavelength due to the polariton (light-exciton coupled mode) effect is confined in the sphere even when its radius is smaller than 100 nm. Such a confined light field forms the whispering gallery mode (WGM)-like spatial structure as shown in Fig. 5.20(b). We call it polaritonic whispering gallery mode (PWGM). This induces strong light-matter coupling and the enhancement of the radiation force and coherent scattering with less absorption even for small nanoparticle (arrow ⟨i⟩ in Fig. 5.17).

Furthermore, in the frequency region between $\hbar\omega_T$ and $\hbar\omega_L$, the incident field is reflected at the surface and cannot propagate in the sphere as shown in Figs. 5.20(d) and (e), which is the nature of the surface-mode. The rate of the light scattering is much larger than the absorption when such a surface-mode-like spatial structure is formed (arrows ⟨iii⟩ and ⟨iv⟩ in Fig. 5.17).

On the other hand, when the electric field is enhanced near the center of the sphere, the ratio of the absorption cross-section to the scattering one is relatively large because the energy of the field cannot dissipate to the outside the sphere efficiently

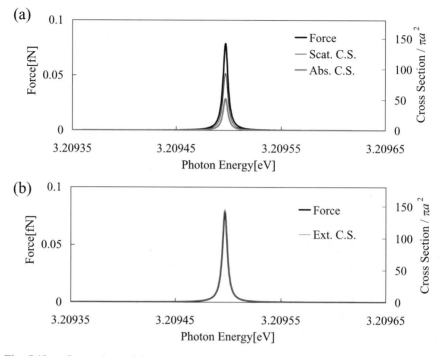

Fig. 5.19. a Comparison of frequency dependencies of scattering cross-section, absorption cross-section and radiation force: Radius $= 10\,[\mathrm{nm}]$; $\gamma = 1.0\,[\mu\mathrm{eV}]$. **b** Comparison of frequency dependencies of extinction cross-section and radiation force: Radius $= 10\,[\mathrm{nm}]$; $\gamma = 1.0\,[\mu\mathrm{eV}]$

(Figs. 5.20(c) and (f)). This is the reason why the absorbing force is dominant at corresponding peak energies (arrow ⟨ii⟩ in Fig. 5.17).

More interesting information obtained from these figures concerns the size dependence of the force. Reflecting the spatial distribution of the electric field, the radiation force by Mie resonance due to the background dielectric is proportional to the surface area of the sphere $4\pi a^2$, whereas the force by excitonic resonance becomes proportional to a^3 as the radius decreases because excitons couple with light over the whole volume of the sphere, as shown in Fig. 5.20(f).

5.3.3 Proposal of Size-Selective Manipulation

In the previous section, we saw that the resonance effect greatly helps to enhance the radiation force exerted on nano objects and that there are conditions wherein the optical process dominated by the coherent scattering with small absorption effectively works for optical manipulation. However, the more important aspect of using the resonance optical process lies in the fact that it can be linked with the quantum mechanical properties of confined electronic systems. This implies that the resonant

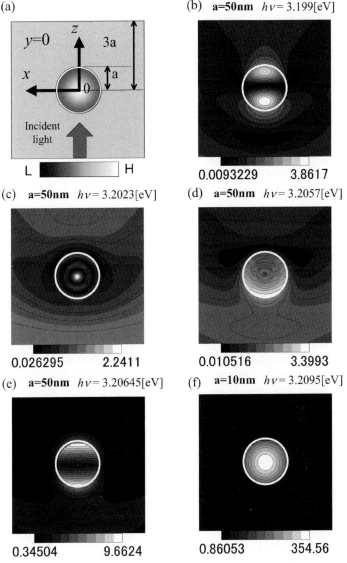

Fig. 5.20. a Geometry of the incident light and a sphere. The sphere is enclosed within a white circle hereinafter. **b–f**: Spatial distribution of the absolute value of the electric field. Resonant: **b** Radius = 50 [nm], photon energy = 3.199 [eV]. **c** Radius = 50 [nm]; photon energy = 3.2023 [eV]. **d** Radius = 50 [nm]; photon energy = 3.2057 [eV]. **e** Radius = 50 [nm]; photon energy = 3.20645 [eV]. **f** Radius = 10 [nm]; photon energy = 3.2095 eV

optical manipulation can select and sort nano objects of particular properties by accessing quantum mechanical individualities of the targets.

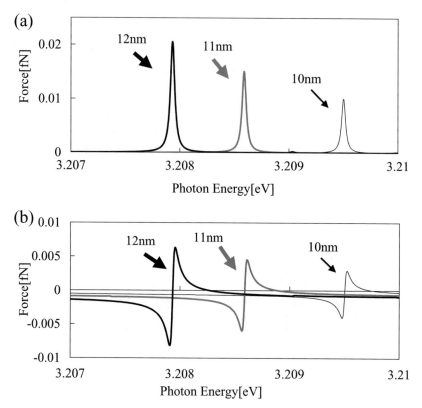

Fig. 5.21. a Frequency dependence of the radiation force on a spherical particle irradiated by a propagating plane wave, where the radius of the sphere is near 10 nm. Size-sensitive shift of the peak of the force is shown in the spectra. **b** Frequency dependence of the radiation force on a spherical particle in a standing wave field, where the radius of the sphere is near 10 nm. Size-sensitive shift of the peak of the force is shown in the spectra

In order to discuss this possibility, we investigate the dependence of the radiation force on the radius of the target sphere in the nanometer range. The quantized excitonic levels are very sensitive to the nanoscale-size changes in the radius, and peaks of the radiation force corresponding to these levels show a similar behavior for both cases of a single plane wave irradiation and a standing wave field irradiation as shown in Figs. 5.21(a) and (b).

By using this size-sensitive shift of the peaks in the force spectra, we can sort out nanoparticles and arrange them by using the periodic potential formed by a standing wave. This effect can be used even for nanoparticle coated or doped with resonance-matter. It is expected that these effects can be applied for the fabrication of nanostructures with high-performance photo-functions, such as photonic crystals including electronic resonance effects.

As shown in Figs. 5.22 and 5.23, the spectral peak position of the force shows less size-sensitive shift at the energy of PWGM in the case that the resonance state

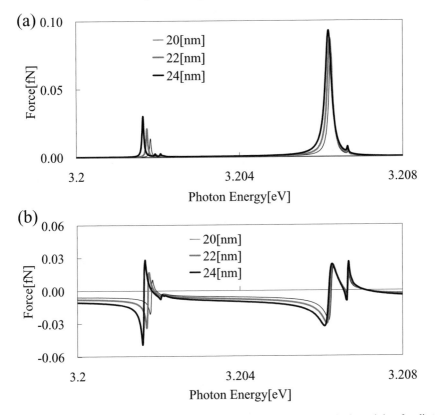

Fig. 5.22. a Frequency dependence of the radiation force when a spherical particle of radius near 20 nm with a local dielectric function is irradiated by a propagating plane wave. **b** Frequency dependence of the radiation force when a spherical particle of radius near 20 nm with a local dielectric function is in a standing wave field

of a sphere is represented by a single Lorentz oscillator, namely, using the dielectric function with infinite effective mass limit. However, even in this case, size selection might be possible depending on damping, the linewidth of the laser and environmental conditions.

5.4 Theoretical Proposal of Nano-Optical Chromatography in Superfluid He^4

In the previous sections, we theoretically investigated the radiation force on a nano object and clarified the presence of various effects useful for optical manipulation under an excitonic resonance condition. Now, we consider the problem of how much effect can be expected under the actual experimental conditions. Although the electronic resonance effect appears at room temperature where nano objects can move in

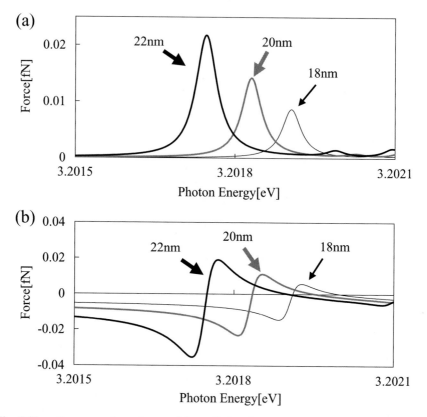

Fig. 5.23. a Frequency dependence of the radiation force when a spherical particle of radius near 20 nm with a local dielectric function is irradiated by a propagating plane wave. **b** Frequency dependence of the radiation force when a spherical particle of radius near 20 nm with a local dielectric function is in a standing wave field. Enlarged view near the energy of the polaritonic whispering gallery mode (PWGM) region

a fluid medium, it is interesting to see the maximum potential of resonant optical manipulation in the first step of the study. It is known that the excitonic resonance effect becomes maximum at cryogenic temperatures. Therefore, here we consider the optical manipulation in superfluid He^4. In a realistic situation, there is inhomogeneity of the resonance width due to nonradiative damping. As for the resonance with wider width, the peak value of the force becomes lower. However, if we apply a laser beam with a line width covering the resonance width of an object, the net radiation force is determined by the overlap integral. In order to take account of this effect, we consider finite line width of the incident laser in the following numerical demonstrations. On the other hand, if we use a beam with sharp line width, the objects with closer resonance to the laser frequency move to a further position. Namely, we can separate the objects with particular resonance as in the case of chromatography if the migration length is macroscopic. How we can clearly separate the target objects depends on the

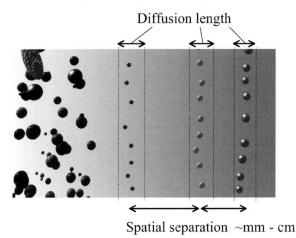

Spatial separation ~mm - cm

Fig. 5.24. Schematic illustration of the macroscopic spatial sorting of nanoparticle with different size by radiation force beyond diffusion length in the superfluid He^4

diffusion length of nano objects. (See schematic image in Fig. 5.24.) In this section, we calculate the radiation force considering the above-mentioned factors to discuss the feasibility of nano-optical chromatography in realistic experimental conditions [63].

5.4.1 For a Laser with Finite Line Width

In this study, we consider a spherical nanoparticle irradiated by a plane wave propagating in the z-direction. The distribution of the incident intensity in the frequency domain is assumed to be $P_{las} w(\omega, \omega_{las})$, where ω_{las} is the center frequency of the laser, P_{las} is the integrated laser intensity, and $w(\omega, \omega_{las})$ is the weighting function. The net radiation force has only a z-component and is described by the following overlap integral as

$$\bar{F}_z(\omega_{las}) = \frac{P_{las} \int d\omega \langle \tilde{F}_z \rangle(\omega) w(\omega, \omega_{las})}{\int d\omega w(\omega, \omega_{las})}, \tag{5.104}$$

where $\langle \tilde{F}_z \rangle(\omega) = \langle F_z \rangle(\omega)/P_{las}$ is the force normalized by P_{las}. The calculation method and the material parameters are the same as in the previous section, namely, we use the ABC method and consider a CuCl Z_3 exciton.

Now, we consider the CW laser light whose frequency distribution follows a Gaussian function as

$$w(\omega, \omega_{las}) = \frac{1}{\sqrt{2\pi}\sigma_G} \exp\left[-\frac{(\omega - \omega_{las})^2}{2\sigma_G^2}\right], \tag{5.105}$$

where the full width at half maximum (FWHM) is $\Gamma_G = \sqrt{8\ln 2}\sigma_G$ (an example of frequency dependence is shown in Fig. 5.25). Laser intensity is assumed to be $P_{las} = 50 \, [\mu W/100 \, \mu m^2]$ that is in the linear response regime as heretofore.

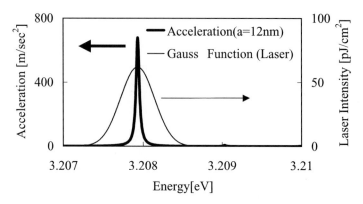

Fig. 5.25. Frequency dependence of the laser power (Gauss function with the center frequency 3.2078 eV): $\Gamma_G = 0.5$ [meV]. Acceleration induced by the radiation force on a spherical particle ($a = 12$ nm) is also shown, where a monochromatic incident light is assumed. Nonradiative damping constant is $\gamma = 0.02$ [meV]

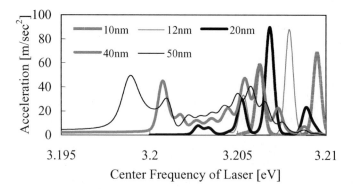

Fig. 5.26. Acceleration induced by the radiation force on a particle is plotted for various radii as a function of the center frequency of incident laser of 0.5 meV width

In Fig. 5.26, for various radii, we calculate the acceleration induced by the radiation force on a particle as a function of the center frequency of the incident light. This result shows that the radiation force is selectively exerted on the particles in a particular size range of nanoscale order even if the laser has a certain width ($\Gamma_G = 0.5$ meV in this calculation, for example).

In order to investigate the laser-width dependence of the acceleration, we evaluate the acceleration of a particle ($a = 12$ nm) as a function of the center frequency of the incident light. The laser frequency is considered to be in the energy region near the eigenenergy of the second TM-mode exciton by changing the nonradiative width and the laser width, as shown in Figs. 5.27(a) and (b). Since the radiative width of the second TM-mode exciton is $\Gamma_\lambda \approx 4.0$ [μeV], which is smaller than assumed γ, the force is approximately proportional to the inverse of γ following (5.63) [since Γ_λ and γ are the half widths at half maximum (HWHM), we must multiply them by 2 if

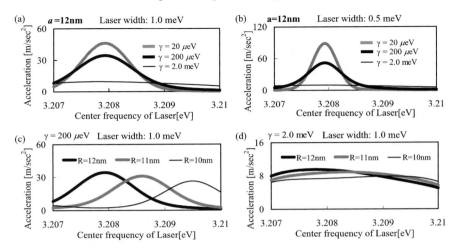

Fig. 5.27. Acceleration induced by the radiation force on a particle of 12-nm radius is plotted as a function of the center frequency of incident laser for each damping constant (γ). **a** Laser width: 1.0 [meV]. **b** Laser width: 0.5 [meV] acceleration induced by radiation force on a particle of radius near 10 nm is plotted as a function of center frequency of incident laser of 1.0 meV width for each nonradiative damping constant γ. **c** $\gamma = 200$ [μeV]. **d** $\gamma = 2.0$ [meV]

we compare these values with Γ_G]. Meanwhile, when the relation $\Gamma_G \gg 2(\Gamma_\lambda + \gamma)$ is satisfied, the radiation force depends not only on Γ_λ and γ but also on Γ_G. As seen in Fig. 5.27(a), by preparing the laser light of width sufficiently wider than $2(\Gamma_\lambda + \gamma)$, we can provide acceleration even to particles with large γ, which is comparable to that of the particles with small γ. Conversely, if we irradiate the particles with a laser light of narrow line width, we can provide large acceleration only to the nanoparticle with small γ by using the light whose center frequency corresponds to the resonance energy (Fig. 5.27(b)). This mechanism can be expected to be useful for sorting out the high-quality nano particles.

5.4.2 Spatial Displacement of Nanoparticles

In order to clarify the possibility of the size-selective nano-manipulation under more realistic conditions, we numerically evaluate the displacement of a nanoparticle by radiation force considering the effects of diffusion and friction in superfluid He^4. For this purpose, we investigate the influence of a finite laser width on the size dependence of a force spectrum. Assuming $\Gamma_G = 1.0$ [meV], we calculate ω_{las}-dependence of the acceleration on particles with 10-, 11- and 12-nm radius. In the case of smaller damping $\gamma = 200$ [μeV] (Fig. 5.27(c)), peaks of acceleration are relatively well separated. On the other hand, in the case of larger damping $\gamma = 2.0$ [meV], which is comparable to Γ_G (Fig. 5.27(d)), the width of each peak becomes very wide and its magnitude decreases, where the size-selective manipulation is seemingly difficult. However, the chromatography-type sorting might be possible even in this case, as shown in the following demonstrations, because the difference of their spatial displacements becomes macroscopic and larger than the diffusion length within the operation time of the manipulation.

The spatial displacement of a particle by the radiation force is evaluated in cases without and with the friction of medium. That in the former case is evaluated by

$$z = z_0 + v_{0z}t + \frac{1}{2}\alpha_z^{(\text{rad})}t^2, \tag{5.106}$$

where the initial position z_0 and initial velocity v_{0z} are assumed to be zero in the latter numerical calculation, $\alpha_z^{(\text{rad})}$ is the z-component of acceleration induced by radiation force and t is the time. That in the latter case is evaluated by the classical Navier–Stokes equation (e.g. [64]) as

$$m_p \ddot{z} = -6\pi \eta a \dot{z} + m_p \alpha_z^{(\text{rad})}, \tag{5.107}$$

where m_p and a are the mass and the radius of nanoparticle, respectively, and η is the viscosity coefficient of the medium (assumed to be $\eta \approx 1.0 \times 10^{-9}$ [Pa·sec] [65]).

We can easily solve (5.107), and according to the solutions, z is proportional to t^2 for $t \ll m_p/6\pi \eta a$, but is proportional to t for $t \gg m_p/6\pi \eta a$.

On the other hand, the diffusion length in the surrounding medium is described as

$$\langle z_{\text{diff}}^2 \rangle = 2D \left[t - \frac{m_p}{6\pi \eta a} \left(1 - \exp \left[-\frac{6\pi \eta a}{m_p} t \right] \right) \right], \tag{5.108}$$

where the diffusion coefficient is described as

$$D = \frac{k_B T}{6\pi \eta a} \frac{\zeta(5/2)}{\zeta(3/2)} \left(\frac{T}{T_c} \right)^{3/2} \tag{5.109}$$

by the Stokes–Einstein relation (superfluid He^4 is treated as Bose gas below the critical temperature $T_c = 2.17$ K) (e.g. [66]) k_B is the Boltzmann factor, T is the temperature (assumed to be 1 K where the superfluid component is dominant (e.g. [67]); $(T/T_c)^{3/2} \approx 0.31283$), and $\zeta(x)$ is the ζ function ($\zeta(5/2)/\zeta(3/2) \approx 0.51378$). Furthermore, in (5.108), $\sqrt{\langle z_{\text{diff}}^2 \rangle}$ is proportional to t for $t \ll m_p/6\pi \eta a$, but is proportional to $\sqrt{t}$ for $t \gg m_p/6\pi \eta a$.

In Fig. 5.28, tuning the center frequency of the incident light to the energy of the second TM mode of the particle of 12-nm radius (3.2078 eV), we calculate the positions of particles of radius near 10 nm as a function of time t, from which we can discover the spatial distribution of respective particles for $\gamma = 200$ [μeV] and $\gamma = 2.0$ [meV]. Even in the case of large damping ($\gamma = 2.0$ [meV]), several tens of milliseconds irradiation of laser light allows us to spatially separate the nanoparticle of different radii over the macroscopic distance greater than the diffusion length of Brownian motion (Fig. 5.28(b)). Of course, in the case of smaller damping ($\gamma = 200$ [μeV]), induced acceleration becomes sufficiently large for the

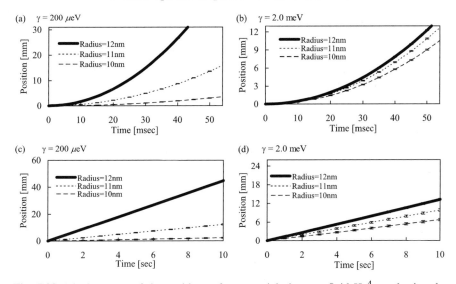

Fig. 5.28. Displacement of the positions of nanoparticle in superfluid He^4, neglecting the friction with the medium, which is evaluated for each nonradiative constant **a** $\gamma = 200$ [μeV] and **b** $\gamma = 2.0$ [meV]. Vertical bars indicate the calculated diffusion length of a particle. Also, displacement of the positions of nanoparticle in superfluid He^4, taking into account the friction with the medium using classical treatment, which is evaluated for each nonradiative constant. **c** $\gamma = 200$ [μeV] and **d** $\gamma = 2.0$ [meV]. Vertical bars indicate the calculated diffusion length of a particle

nanoparticle of different radii to be separated from each other over a long distance, as shown in Fig. 5.28(a). From the comparison of Figs. 5.28(a) and (b), we know that the displacements are much different between the particles with different γ (even with the same radius). On the other hand, as shown in Figs. 5.28(c) and (d), even in the presence of the friction of medium as shown in (5.107), we can perform such a spatial separation over a distance greater than the diffusion length with several seconds irradiation of laser light. This fact is favorable to experimentally observe the nanoparticles changing their positions in the slow time scale. These results indicate that the technique of "nano-optical chromatography" might be possible by using resonant radiation force.

5.5 Experiment of Optical Transport of Nanoparticles

As explained in the introduction, the experiment of optical manipulation of nano objects is rather challenging because the light scattering (absorption), or induced polarization is too weak to yield sufficient force to manipulate nano objects in a usual nonresonance condition. On the other hand, researchers have been cautious about using resonant light considering that the resonance condition causes damage to the sample due to heating. However, recently, some experimental trials have been conducted for optical manipulation using resonant light. In these cases, they carefully use the resonant condition to make use of the potential of resonant radiation force. Hosokawa et al. showed the trapping of dye-doped nanoparticle of organic molecules using a strongly focused laser beam [68]. In this experiment, nonresonant laser beams are mainly applied and a resonant laser is used to assist the trapping of molecules more strongly. The result shows that the use of a resonant laser beam considerably extends the trapping time. Li et al. attempted the resonance trapping of individual multiple fluorophore-labeled antibodies, and they concluded that selective resonance trapping to sort and manipulate fluorophore-labeled biomolecules and complexes might be possible [69].

On the other hand, we have experimentally demonstrated optical transport of a semiconductor nanoparticle to show the potential of resonant optical manipulation as a chromatography-type technique [70]. In order to make the best use of the resonant effect of confined excitons in a nanoparticle, we performed this experiment in superfluid He^4 as proposed in the previous section. The next section briefly explains this experiment.

5.5.1 Introduction of Nanoparticles into Superfluid He^4

In the experiment of optical manipulation of nanoparticles in superfluid He^4, the first challenge to overcome is how to introduce nanoparticles into superfluid He^4. In our experiment, we directly fabricate nanoparticles in superfluid He^4 by means of laser ablation. This technique is an application of laser sputtering developed to introduce atoms into superfluid He^4 [71]. The target is a pressed tablet of CuCl powder. Setting this tablet on the sample holder at the center of the cryostat, it is irradiated with a light pulse from a Q-switched, frequency-tripled Nd:YAG laser at 355 nm (pulse duration $\sim$10 ns, repetition rate $\sim$10 Hz, pulse intensity 0.3 J/cm^2). Analysis using scanning electron microscopy (SEM) and energy dispersive X-ray spectroscopy (EDX), shows that the yielded nanoparticles consist of Cu and Cl (the detail is shown in [70]) and that they have excellent morphology as particles. The size range of observable nanoparticles is approximately from 10 nm to 10 μm. A typical SEM image of the CuCl nanoparticle on a silicon substrate is shown in Fig. 5.29(b). These results indicate that we have successfully fabricated CuCl nanoparticles with a certain crystal quality.

5.5.2 Optical Transport of Nanoparticles Using Resonant Light

In order to demonstrate optical transport, we irradiate floating CuCl nanoparticles with a laser light under an excitonic resonance condition. The second harmonics of a mode-locked Ti:Sapphire laser (pulse duration $\sim$100 fs, repetition rate $\sim$80 MHz, spectral width $\sim$20 meV) were used as a manipulation light. The center energy of laser light is tuned to around 3.2 eV, which corresponds to the resonance energy of Z_3 exciton in the bulk CuCl. The spectral width of the incident laser covers resonance energies of particles of smaller radius than 100 nm. We attempted to transport the nanoparticles horizontally, where they are sorted by the balance between the radiation force and gravity.

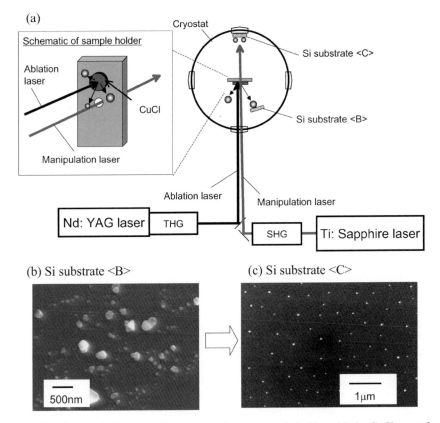

Fig. 5.29. a Schematic diagram of experimental setup: sample holder with the CuCl target for ablation is set at the center of the cryostat. Inset: The manipulation laser passes through the pinhole on the sample holder. **b** SEM image of nanoparticles on silicon substrate obtained by laser ablation in superfluid He^4: particles with a broad size distribution ranging from 10 nm to 10 μm were observed. **c** SEM image of nanoparticles on silicon substrate irradiated by manipulation laser: particles of diameters from 10 nm to 50 nm are observed

The experimental setup is illustrated in Fig. 5.29(a). Both the ablation laser and the manipulation laser are incident from the same window. The manipulation laser passes through the small hole on the sample holder under the ablation laser beam, and free-falling nanoparticles are pushed by the manipulation laser onto the Si substrate located behind the sample holder. The distance between the sample holder and the Si substrate is a few centimeters. The manipulation laser beam was collimated to ~200 μm in diameter with an intensity of 150 W/cm^2. The total duration of the laser irradiation was approximately one hour—that is much longer than the time required to transport a nanoparticle over a few centimeters, as discussed in the previous section.

In this experiment, we use resonant and nonresonant lasers for manipulation, where the photon energy of the latter was 3.01 eV. In the case of the nonresonant laser, nanoparticles were not found on the Si substrate. On the other hand, in the case of resonant laser irradiation, we found a lot of nanoparticles with a radius from 10 nm to 50 nm adhered to the substrate as seen in Fig. 5.29(c). Namely, the nanoparticles within a particular size range were selectively transported over a macroscopic distance by resonant radiation force. Moreover, in the SEM image, we find that the transported particles are placed at certain intervals. Although the mechanism of this phenomenon is still unclear, we speculate that some kind of light-induced force between them prevents their aggregation. In [52], we discussed the radiation-induced interparticle force, which would be one of the possible mechanisms.

In the theoretical estimation, it is considered that the acceleration provided to CuCl nanoparticles having radius of several tens of nanometers becomes several hundred times larger than the gravitational one in the presence of excitonic resonance. In addition, the spectral peaks of acceleration under the resonance condition are mainly concentrated in the energy region (3.19–3.21 eV) covered by manipulation laser used in the experiment. Therefore, we consider that nanoparticles in Fig. 5.29(c) have been transported by the resonant radiation force much stronger than the gravitational force. If we use the laser light with sharper line width than that of the present laser with a broad spectral width, it would be possible to realize more precise size selection as proposed in Sect. 5.3.

Although extracting quantitative information on the resonant radiation force from a similar experiment is our future subject, the present result clearly shows the potential of the resonant optical manipulation, which strongly encourages further experimental studies to develop new types of optical manipulation linked to the world of quantum mechanics in the nanoscale regime.

5.6 Summary and Future Prospects

In this article, we introduced our theoretical studies on the "mechanical interaction between nano object and electronic resonant light" and the recent experimental study of optical manipulation of nanoparticles with excitonic resonance. From the results of the theoretical studies, we proposed a new type of optical manipulation linked with the quantum mechanical properties of nano objects. The realizability of this proposal was demonstrated by the experiment of optical transport of nanoparticles in superfluid He^4. In this section, we summarize the results of these studies and provide future prospects.

5.6.0.1 Establishment of the Theoretical Method Applicable to the System "from Atom to Macroscopic Object" and the General Properties of Radiation Force in Each Size Regime

We derived the unified analytical expression of the radiation force on the system "from atom to macroscopic object" based on the Lorentz force equation and microscopic nonlocal theory. In order to understand the general properties of the "mechanical interaction between nano object and electronic resonant light", we applied this formula to several simple models and obtained the explicit expressions of the force for respective models. These discussions help us to give physical explanations of the results obtained by the numerical calculations demonstrated in the latter parts.

The obtained expressions reproduce all the properties explained by the conventional expressions of radiation force from the classical regime to the atomic regime. Furthermore, these expressions include generalizations corresponding to the features specific to nanoscale systems reflecting the spatial extension of electronic wavefunctions. In addition, we clarified the existence of the new types of additional terms, other than the formally known scattering force, absorbing force and gradient force. These terms arise from the asymmetric spatial structure of the internal field reflecting the microscopic spatial structure of the coupled modes of the light and the confined electronic systems.

5.6.0.2 The Mechanism of the Radiation Force on a Single Nano Object and Feasibility of the Experiment

We investigated the mechanism of the radiation force on a single semiconductor spherical nanoparticle by numerical calculations. From the obtained results, we clarified the presence of various favorable effects in the resonant optical manipulation of nano objects as follows:

(1) As the size decreases, the merit in using the excitonic resonance greatly increases. The enhancement of the force by using electronic resonance is remarkable for a particle with a radius of less than 100 nm for semiconductor materials such as CuCl. For spherical particles of around 10-nm radius, the force near the excitonic resonance can be especially enhanced as much as four or five orders of magnitude larger than that in the absence of the resonance (in the calculation, nonradiative damping constant is assumed to be 20 [μeV], which can be obtained in cryogenic conditions, such as superfluid He^4).

(2) We can find the condition wherein the heat accumulation can be avoided even when using an electronic resonant light. For a particle having the radius of several tens of nanometers, the radiative scattering can be dominant depending on the frequency due to the strong radiation-matter coupling arising from the coherent extension of the excitonic center-of-mass wavefunction over the whole volume. This mechanism is favorable for some kind of optical manipulation under the resonance condition.

(3) The spectral peak position of the exerted force is considerably sensitive to nanoscale-size changes because of the large shift of the quantized levels including the radiative shift, which would be useful for highly accurate size-selective manipulation.

In addition, assuming more realistic conditions, we discussed the feasibility of the experiment of the manipulation of excitonic-active nano objects considering the effects of the width of incident laser and the surrounding medium. We considered superfluid He^4 as a surrounding medium, where the cryogenic condition is realized and the viscosity is almost zero. We evaluated the moving distance by the radiation force induced by the incident light with a finite width and diffusion length in super-fluid He^4 as a function of time. As a result, we clarified that particular nanoparticles, whose resonance energy corresponds to the center frequency of the incident laser, move away from others over a macroscopic distance. This distance is much longer than the diffusion length, even in the presence of the diffusion and the friction of the superfluid He^4. Although it is preferred that the damping constants of nanoparticles are small and the width of the incident laser is narrow in such a spatial separation of nano objects, we can realize that the spatial separation of particular nanoparticle even when the homogeneous widths of nanoparticle are large if the width of the incident laser is narrower than or comparable to them. This result indicates the possibility to map the size distribution of nano objects onto their macroscopic spatial distribution if we create appropriate conditions with the incident laser light (nano-optical chromatography).

5.6.0.3 Experiment of Resonant Optical Manipulation of Nanoparticles in Superfluid He^4

Optical manipulation of nano objects has been considered to be challenging because the light scattering (absorption) or induced polarization is too weak to yield sufficient force to manipulate nano objects in usual nonresonance conditions. However, if we use the electronic resonance effect in an appropriate way, a sufficiently strong force for manipulation can be exerted avoiding the heating problem. Further, using the resonance effect in nano objects links the radiation force to the quantum mechanical properties of nanoscale materials, which would lead to quite a new type of optical manipulation to sort quantum mechanical individualities of nano objects.

The fundamental possibility of such a theoretical proposal was successfully demonstrated by the experiment. In order to reveal the maximum potential of the resonant optical manipulation, we performed the experiment in a cryogenic environment where the excitonic resonance effect optimally appears. Performing the optical transport of CuCl nanoparticles, which are created by laser ablation method directly in superfluid He^4, over a macroscopic distance, we have clarified the following: (1) A resonant laser light exerts sufficiently strong force to convey nanoparticle with a radius of several tens of nanometers in superfluid He^4 over a macroscopic distance, whereas a nonresonant laser does not. (2) Due to the balance between the gravity and radiation force, nano objects in a particular size range are selectively transported.

Although a quantitative evaluation of the experimental results and a study of the possibility of nanoscale size selection are our future subjects, the present results have opened the door to a new research field of optical manipulation and to novel technologies for the fabrication of nanostructures, strongly encouraging further experimental studies to develop new type of optical manipulation linked to the quantum mechanics in the nanoscale regime.

5.6.0.4 Future Prospects

Since the success of the experimental demonstration of nano optical manipulation using resonant light, a new research phase has started where we should investigate quantitative aspects of the experiment by comparing it with theoretical studies in detail. What is the magnitude of the force exerted on a nanoparticle in the experiment in superfluid He^4? What is the influence of the nonradiative damping of nanoparticles? What is the exact size distribution of the transported particles? These are some of the questions that we will study in this research phase.

The experimental trials for the various types of laser beam are important subjects for the future. The studies on the trapping of nano objects by using a strongly focused beam, their positional control using a standing wave, the control of rotating motion by a wavefront-controlled beam such as a Laguerre-Gaussian beam are desired to develop various types of motion-control techniques. Further, studies on the techniques that use multiple light sources—for example, the combination of a propagating field and an evanescent one, the combination of beams with different frequencies and so on—are necessary to extend the degree of freedom of optical manipulation using resonant light.

The experiment introduced in the previous section was performed in a particular environment, namely, at a cryogenic temperature to demonstrate the maximum performance of resonant manipulation. However, it would be possible that resonant optical manipulation is performed in other environments, including a room-temperature environment, as molecular trapping in [68, 69]. Therefore, more experimental trials in various kinds of environments are desired.

If the technique of the resonant optical manipulation is established, it would be an important breakthrough to develop novel fabrication technology to realize ultra-high homogeneity of materials. In usual fabrication technologies to control structural parameters such as size, shape and position, certain inhomogeneity of materials cannot be avoided. Further, the homogeneity of the material with respect to a single kind of parameter does not necessarily guarantee the improvement of the device function. For example, even if we can make the size distribution of the particles very small, it does not directly lead to high homogeneity of the resonance line because it is not determined only by the particle size. However, the resonant optical manipulation directly accesses the quantum mechanical properties such as the electronic resonance level. Namely, this technique makes it possible to directly control the homogeneity of the resonance line itself, which could be a novel approach to the breakthrough of nano fabrication technology. If this type of technology is realized, it would lead to the new device and material technologies, where the enormous improvement of quantum dot devices—such as a quantum dot laser, a high-performance single photon source with quantum dots, and a high-performance biomarker with extremely sensitive wavelength selectivity—would be realized. We strongly hope that our study of resonant optical manipulation leads to such novel technologies in the future.

Acknowledgments

The authors are grateful to Professor K. Cho for his continued inspiration through fruitful discussions and encouragement. They also thank Professor H. Masuhara, Dr. H. Ajiki, and Professor K. Kobayashi for their useful discussions. The main part of the experiment of resonant optical manipulation in superfluid He^4 was performed by Professor T. Itoh's group at Osaka University. The authors also thank the members of his group, Professor T. Itoh, Professor M. Ashida, and Mr. K. Inaba for their close collaboration and fruitful discussions. A great part of the research reported in this article was supported by CREST program of Japan Science and Technology Agency.

References

[1] O. Marti, V.I. Balykin, Light forces on dielectric particles and atoms, in *Near Field Optics*, ed. by D.W. Pohl, D. Courjon. NATO ASI Series (Kluwer Academic, Dordrecht, 1993)

[2] D.G. Grier, Nature **424**, 810 (2003)

[3] P. Lebedev, Ann. Phys. (Leipzig) **6**, 433 (1901)

[4] E.F. Nicholls, G.F. Hull, Phys. Rev. **13**, 207 (1901)

[5] A. Ashkin, Phys. Rev. Lett. **25**, 1321 (1970)

[6] C. Cohen-Tanoudji, in *Fundamental Systems in Quantum Optics*, Proceedings of the Les Houches Summer School, Session L III, ed. by J. Dalibard, J. Raimond, J. Zinn-Justion (North-Holland, Amsterdam, 1992)

[7] C. Cohen-Tanoudji, J. Dupont-Roc, G. Grynberg, *Atom–Photon Interaction*, Wiley Science Paperback Series (Wiley, New York, 1992)

[8] S. Chu, L. Hollberg, J.E. Bjorkholm, A. Cable, A. Ashkin, Phys. Rev. Lett. **55**, 48 (1985)

[9] S. Chu, L. Hollberg, J.E. Bjorkholm, A. Cable, A. Ashkin, Phys. Rev. Lett. **57**, 314 (1986)

[10] W.D. Phillips, J.V. Prodan, H.J. Metcalf, J. Opt. Soc. Am. B **2**, 1761 (1985)

[11] M.H. Anderson, J.R. Ensher, M.R. Matthews, C.E. Wieman, E.A. Cornell, Science **269**, 198 (1995)

[12] K.B. Davis, M.-O. Mewes, M.R. Andrews, N.J. van Druten, D.S. Durfee, D.M. Kurn, W. Ketterle, Phys. Rev. Lett. **75**, 3969 (1995)

[13] A. Ashkin, Phys. Rev. Lett. **24**, 156 (1970)

[14] A. Ashkin, J.M. Dziedzic, J.E. Bjorkholm, S. Chu, Opt. Lett. **11**, 288 (1986)

[15] M.M. Burns, J.-M. Fouriner, J.A. Golvchenko, Science **249**, 749 (1990)

[16] T.T. Perkins, D.E. Smith, S. Chu, Science **264**, 819 (1994)

[17] P. Galajda, P. Ormos, Appl. Phys. Lett. **78**, 249 (2001)

[18] S. Juodkazis, N. Mukai, R. Wakai, A. Yamaguchi, S. Matso, H. Misawa, Nature **249**, 178 (2000)

[19] H.-J. Güntherodt, D. Anselmetti, E. Meyer (eds.), *Forces in Scanning Probe Method*, NATO ASI Series (Kluwer Academic, Dordrecht, 1995)

[20] H. Masuhara, H. Nakanishi, K. Sasaki (eds.), *Single Organic Nanoparticles* (Springer, Berlin, 2002)

[21] K. Sasaki, M. Tsukima, H. Masuhara, Appl. Phys. Lett. **71**, 37 (1997)

[22] K. Svoboda, S.M. Block, Opt. Lett. **19**, 930 (1994)

[23] T. Sugiura, T. Okada, Y. Inoue, O. Nakamura, S. Kawata, Opt. Lett. **22**, 1663 (1997)

[24] K. Sasaki, J. Hotta, K. Wada, H. Masuhara, Opt. Lett. **25**, 1385 (2000)
[25] S. Ito, H. Yoshikawa, H. Masuhara, Appl. Phys. Lett. **78**, 2566 (2001)
[26] P. Debye, Ann. Phys. **30**, 57 (1909)
[27] H.C. van de Hulst, *Light Scattering by Small Particles* (Dover, New York, 1981)
[28] G.F. Bohren, D.R. Huffman, *Absorption and Scattering of Light by Small Particles* (Wiley Interscience, New York, 1983)
[29] J.P. Gordon, Phys. Rev. A **8**, 14 (1973)
[30] J.P. Barton, D.R. Alexander, S.A. Schaub, J. Appl. Phys. **66**, 4594 (1989)
[31] T. Sugiura, S. Kawata, Bioimaging **1**, 1 (1993)
[32] T. Tlusty, A. Meller, R. Bar-Ziv, Phys. Rev. Lett. **81**, 1738 (1998)
[33] A. Ashkin, Biophys. J. **61**, 569 (1992)
[34] K.S. Yee, IEEE Trans. Antennas Propag. **14**, 302 (1966)
[35] L. Novotny, R.X. Bian, X.S. Xie, Phys. Rev. Lett. **79**, 645 (1997)
[36] K. Okamoto, S. Kawata, Phys. Rev. Lett. **83**, 4534 (1999)
[37] E.M. Purcell, C.R. Pennypacker, Astrophys. J. **186**, 705 (1973)
[38] P.C. Chaumet, M.N. Vesperinas, Phys. Rev. B **62**(11), 185 (2000)
[39] P.C. Chaumet, M.N. Vesperinas, Phys. Rev. B **64**, 035422 (2001)
[40] P.C. Chaumet, A. Rahmani, M.N. Vesperinas, Phys. Rev. Lett. **88**, 123601 (2002)
[41] P.C. Chaumet, A. Rahmani, M.N. Vesperinas, Phys. Rev. B **66**, 195405 (2002)
[42] R.R. Agayan, F. Gittes, R. Kopelman, C.F. Schmidt, Appl. Opt. **41**, 2318 (2002)
[43] H. Ishihara, K. Cho, Phys. Rev. B **53**, 15823 (1996)
[44] H. Ishihara, K. Cho, K. Akiyama, N. Tomita, Y. Nomura, T. Isu, Phys. Rev. Lett. **89**, 017402 (2002)
[45] K. Cho, *Optical Response of Nanostructures* (Springer, Berlin, 2003)
[46] K. Cho, J. Phys. Soc. Jpn. **55**, 4113 (1986)
[47] K. Cho, Prog. Theor. Phys. Suppl. **106**, 225 (1991)
[48] J.D. Jackson, *Classical Electrodynamics*, 3rd ed. (Willey, New York, 1999)
[49] W.C. Chew, in *Waves and Fields in Inhomogeneous Media* (Van Nostrand Reinhold, New York, 1990)
[50] P.M. Morse, H. Feshbach, *Method of Theoretical Physics, Part II* (McGraw-Hill, New York, 1953)
[51] T. Iida, H. Ishihara, J. Lumin. **108**, 351 (2004)
[52] T. Iida, H. Ishihara, Phys. Rev. Lett. **97**, 117402 (2006)
[53] T. Iida, H. Ishihara, Nanotechnology **18**, 084018 (2007)
[54] H. Ishihara, K. Cho, Optical manipulation of single nanoparticles, in *Single Organic Nanoparticles*, ed. by H. Masuhara, K. Sasaki (Springer, Berlin, 2003), p. 147
[55] T. Iida, H. Ishihara, Phys. Rev. Lett. **90**, 057403 (2003)
[56] T. Iida, H. Ishihara, Phys. Status Solidi, B **238**, 241 (2003)
[57] S.I. Pekar, Sov. Phys. JETP **6**, 785 (1957)
[58] L. Birman, in *Excitons*, ed. by E.I. Rashba, M.D. Sturge (North-Holland, Amsterdam, 1982), p. 27
[59] R. Ruppin, J. Phys. Chem. Solids **50**, 877 (1989)
[60] H. Ajiki, T. Tsuji, K. Kawano, K. Cho, Phys. Rev. B **66**, 245322 (2002)
[61] T. Mita, N. Nagasawa, Solid State Commun. **44**, 1003 (1982)
[62] M. Ikezawa, Y. Masumoto, Phys. Rev. B **61**, 12662 (2000)
[63] T. Iida, H. Ishihara, IEICE Trans. Electron. **88-C**, 1809 (2005)
[64] T.E. Fabor, *Fluid Dynamics for Physists* (Cambridge University Press, New York, 1995)
[65] P. Kapitza, Nature **141**, 74 (1938)
[66] A. Borodin, P. Salminen, *Handbook of Brownian Motion: Facts and Formulae* (Birkhäuser, Boston, 1996)

[67] R.J. Donnelly, *Experimental Superfluidity* (University of Chicago Press, Chicago, 1967)
[68] C. Hosokawa, H. Yoshikawa, H. Masuhara, Jpn. J. Appl. Phys. **45**, L453 (2006)
[69] H. Li, D. Zhou, H. Browne, D. Klenerman, J. Am. Chem. Soc. **128**, 5711 (2006)
[70] K. Inaba, K. Imaizumi, K. Katayama, M. Ichimiya, M. Ashida, T. Iida, H. Ishihara, T. Itoh, Phys. Status Solidi B **243**, 3829 (2006)
[71] A. Fujisaki, K. Sano, T. Kinoshita, Y. Takahashi, T. Yabuzaki, Phys. Rev. Lett. **71**, 1039 (1993)

Index

α-cyano-4-hydroxycinnamic acid (CHCA)
    68
2,5-dihydroxybenzoic acid (DHB) 68, 92
2D electron gas 25, 27, 29

ABC 132, 133, 135, 155
absorbing force 117, 130, 143, 145, 147,
    148, 150, 163
absorption cross-section 143, 145, 147–150
additional boundary condition 132
adiabatic approximation 42, 46
alkali metals attachment 68
angiotensin I 84
anodic oxidation 74, 77, 78
anodization voltage 74, 77, 78
anti-Hermitian operator 54, 55
aperture–NSOM probe 3, 4
argon ion sputtering 67, 73–75
argon laser 90

biexciton 13, 16, 17, 22–26
bilayer resist process 100, 104
bradykinin 82, 83, 89, 90, 93, 94
Brownian motion 116, 121, 158

carrier localization 31, 32
chemical etching 4, 78, 85, 86
chlorauric acid 77
citric acid 82, 88, 89, 94
coarse-grained 43, 48
coherent state 56, 57, 61, 63, 64
commutation relation 50, 54, 55
continuity equation 126
coupling constant 48, 53, 54, 57
CuCl 134, 136, 155, 160–164

Davydov transformation 52, 54, 56, 61, 64
DDA 119
defect 48, 51

delayed extraction 70, 80
delocalization 31, 48, 64
delocalized 2, 13, 16, 32, 34–37, 48, 51, 57,
    60, 64
delocalized phonon 52
desorption/ionization on silicon (DIOS) 69
dilute nitride semiconductor 31
dipole-forbidden 41
discrete dipole approximation 119
dissipative force 119, 130, 131, 143
dressed photon 55, 56, 63, 64

effective interaction 42, 43
electro-deposition 73, 77, 78
electrochemical etching 67, 74
electromagnetic field enhancement 71, 95
electronic excitation 42, 45, 47, 52, 53, 63,
    64
EPP model 44, 46
etching time 76, 85, 86, 105
excitation probability 57
excitation transfer 41
exciton 1, 13, 14, 16–26, 33–36, 46, 47, 121,
    132–135, 147, 149, 155, 156, 160
exciton excited state 22, 25
exciton pooling 69, 90
excitonic center-of-mass motion 132
extinction cross-section 143, 145, 147–150

far-field excitation 19, 20
FDTD 119
field induced quantum dot 26
finite-difference time-domain method 6, 107,
    119
Franck–Condon principle 42
frequency-doubled Nd:YAG laser 67, 79, 80,
    91
frequency-tripled Nd:YAG laser 68, 82

Springer Series in
OPTICAL SCIENCES

Volume 1

Published titles since volume 110

Springer Series in
OPTICAL SCIENCES